Pro Tools 8 Kit

Pro Tools 8 Kit

The Complete Professional Workflow for Music Production

Robert J.Shimonski
Chris Basile

Amsterdam • Boston • Heidelberg • London • New York • Oxford • Paris
San Diego • San Francisco • Singapore • Sydney • Tokyo

Focal Press is an imprint of Elsevier

Focal Press is an imprint of Elsevier
30 Corporate Drive, Suite 400, Burlington, MA 01803, USA
Linacre House, Jordan Hill, Oxford OX2 8DP, UK

∞ Recognizing the importance of preserving what has been written, Elsevier prints its books on acid-free paper whenever possible.

Library of Congress Cataloging-in-Publication Data

Shimonski, Robert.
Pro Tools 8 kit : the complete professional workflow for music production / Robert J. Shimonski ; Chris Basile.
p. cm.
Includes index.
ISBN 978-0-240-81115-4 (pbk. : alk. paper)
1. Pro Tools. 2. Digital audio editors. I. Basile, Chris. II. Title.
ML74.4.P76S52 2009
781.3'4536--dc22

2009018379

British Library Cataloguing-in-Publication Data

A catalogue record for this book is available from the British Library.

ISBN: 978-0-240-81115-4

For information on all Focal Press publications
visit our website at www.books.elsevier.com

Typeset by: diacriTech, Chennai, India

09 10 11 12 13 5 4 3 2 1

Printed in the United States of America

Contents

Preface

Welcome to *Pro Tools 8 Kit – The Complete Professional Workflow for Music Production*, brought to you by Focal Press! In this book and on the companion Web site, we will have a detailed look at all of the steps you will need to use Pro Tools 8 LE effectively in a production environment. You will find a step-by-step process not only to get you up to speed with Pro Tools 8 LE but also to become efficient with its use, so that you can record and produce audio in a professional recording studio or in your own home-based project studio. This book is full of helpful hints on how to quickly navigate Pro Tools 8 and take advantage of its offerings. We explore a large amount of production-specific problems and their solutions throughout the book, providing answers to commonly asked questions.

Today, much of the world relies on digital audio to get news or education or to communicate. The production of a podcast, ringtone, loop, sample, or full-length album are only a handful of examples showing how much we use and rely on digital audio today. When producing this audio, a complete workflow for music production would be one that starts with a solid plan and ends with a deliverable that exceeds your expectations, and this is no easy task. There is much to do, such as preproduction, the recording process, editing, mixing, mastering, and finalizing your work. There are also steps taken to ensure a backup of that work is produced for safe-keeping. Obviously, there is much more to do, such as setting up the equipment or software needed and the actual recording process itself. It's important to learn and master this entire process so that as an audio engineer or producer you can work in a productive and efficient manner to produce the best quality work possible. You also do not want to spend weeks behind the console when time is a factor.

In this book and on the Web site, we cover each step you need to take to produce top-notch-quality audio and give literally hundreds of tips along the way to make your time spent behind the console easier and more productive. This book covers new functionality found in Pro Tools 8, such as new scoring and composing tools, using Sibelius, and working with new session templates that inevitably will make your life easier during the setup phase when starting a recording. In addition, we also cover exploring your Digital Audio Workstation (DAW), setting up Pro Tools, treating your area, using digital plug-ins, instruments, microphones, outboard gear, and much more. You will find many helpful solutions to help maximize your effectiveness while producing and engineering – especially in long sessions where being effective on a control surface or keyboard can save you time, money, and, in some cases, injury. This book was designed to help you get started with Pro Tools 8 LE and by the time you complete the kit feel confident that you will have mastered all the steps needed to work with and record with Pro Tools 8 to produce a product for distribution.

We will also illustrate methods for successfully recording and editing as a professional and/or project studio producer (including information on some useful additional equipment) to guarantee a smooth running production session and give you the tools necessary to produce content with Pro Tools 8 LE. The book is broken down into seven chapters to cover each step of the production process:

Chapter 1 – Introduction

In this chapter, we start with a look at how to prepare a DAW for use and how to set up all needed hardware before installing Pro Tools 8. This section covers all current Digidesign hardware offerings that interoperate with Pro Tools 8 LE, how to select which ones you need, how to configure and connect them properly, and any other preinstallation steps required before performing an installation of Pro Tools 8 LE. Additionally, we will discuss steps to soundproof a space, as well as headphones and monitor/speaker selection. This chapter also covers how to prepare for and perform an upgrade of Pro Tools 8 LE on an Apple Macintosh system, use the iLok key, and much more.

Chapter 2 – Session setup

Now that your DAW is complete, you are ready to start using Pro Tools 8 LE and taking advantage of its many features. In this chapter, we will start up a new session and learn about session parameters, using common tools, and navigating the workspace. We will also learn how to use session templates, the Mix and Edit windows, Transport tools, tracks, plug-ins, routing, I/O, and preferences. This chapter is a complete walkthrough and gives you a great understanding of how to work with many of Pro Tools' most critical features. It will get you ready to begin the next phase of music production – composing your work!

Chapter 3 – Composing

Now that you are familiar with Pro Tools, let's get to work. In this chapter, we will look at how Pro Tools 8 LE allows you to score and compose like a true professional. With Pro Tools 8 LE, you can now build upon your creativity with new and enhanced tools such as the fully integrated MIDI and Score Editor windows, which allow for more possibilities when editing and working with MIDI. This chapter covers scoring, composition, and the use of MIDI devices, as well as the new MIDI functionality in Pro Tools 8 LE and Sibelius.

Chapter 4 – Recording

Now let's get to work recording audio, using microphones and learning about how to track with Pro Tools 8 LE in a production environment. In this chapter, we will look at the recording process in detail, how Pro Tools 8 LE functions in a recording session, and the new functionality that enables you to be more productive when at the console. We will cover the use of recording with microphones, amplifiers, and instruments, including live instruments like drums and percussion. We will also look at the numerous products that now ship with Pro Tools LE such as AIR virtual instruments.

Chapter 5 – Editing

In this chapter, we will take your recorded work and show you how to manipulate tracks and regions and enhance the recording with new tools and features found in Pro Tools 8 LE. We will learn about the Edit window, editing modes, editing tools such as the Zoomer, Smart, and Grabber tools, and much more. Additionally, we will look at how to use fades, work with playlists, and use Beat Detective.

Chapter 6 – Mixing

In this chapter, we will cover how to mix and prep your work for final mastering with Pro Tools 8. We cover mixing concepts, use of effects, automation, and much more. We will also take a close look at not only using digital plug-ins but also outboard gear to enhance your mix and then your final mixdown.

Chapter 7 – Delivery

In this chapter, we cover the final transport of your session, including the session bounce, and how to choose from the various options you can select from. This chapter also covers the steps you take after final mixdown to create a high-quality CD and also discusses how to properly use iTunes and other importing/exporting tools. In this chapter, we also cover how to prepare for any disaster. As you will discover, any production facility today that works with digital audio or video files finds hard disk space to be a challenge. Finally, it covers how to store your sessions and software, prepare for the worst, and get back online if any issues do occur via a backup of your system.

Appendix A – Keyboard shortcuts

The ultimate printout and quick reference guide to mastering the keyboard with Pro Tools 8.

Pro Tools 8 Web site

The Web site has five video modules, taking you through DAW setup, recording, editing, mixing, and delivery. This book and the Web site take the Pro Tools user through every step of the production process and demonstrate exactly what the software can deliver in a production session.

The step-by-step approach allows Pro Tools users to learn new features and achieve professional results fast. There are many ways of working with Pro Tools, and this book is unique in that it demonstrates multiple methods of working, allowing the user to choose the method that works for them and the situation they are in by illustrating each item with clear text- and image-based examples. The goal of this book and the Web site is to serve as a guide and a reference tool to help you become familiar with the recording and production process using Pro Tools 8 LE. This book can help you understand the toys, tools, and day-to-day practice of music recording and production.

We hope that you enjoy this work as much as we enjoyed making it for you. Now, let's get to work!

About the Authors

Rob Shimonski is a well-noted author and educator in the fields of business and technology. An avid studio designer, audio engineer, studio musician, and Pro Tools expert, Rob works in today's most challenging environments providing solutions and delivering results.

Chris Basile is a veteran producer, recording engineer, and musician who specializes in studio design and deployment, Pro Tools LE and HD systems, as well as both audio and video production. He has spent over 20 years on both sides of the glass and just recently opened SubSonic Audio, a full-service recording facility in New York.

Both Rob and Chris are specialists in DAW design and deployment and helped to test Pro Tools 8 LE with Digidesign prior to its official release. Rob and Chris can be found online at http://www.protools101.com, where you can learn more about the authors and the book, and download materials, sample sessions, podcasts, videos, tips, and white papers.

Acknowledgments

Rob and Chris would like to thank Catharine Steers, Carlin Reagan, David Bowers and Monica Mendoza from Focal Press for their help in creating the Pro Tools 8 Kit, and Greg Robles at Digidesign for his help with the Pro Tools 8 beta program. A special thanks is due to to Bruce Bartlett for his help in technically editing the work.

In this chapter

1 Introduction

In this chapter, we start with a look at how to prepare a Digital Audio Workstation (DAW) for use and how to set up all needed hardware before installing Pro Tools 8. This section covers all current Digidesign hardware offerings that interoperate with Pro Tools 8 LE, how to select which ones you need, how to configure and connect them properly, and any other preinstallation steps required before performing an installation of Pro Tools 8 LE. Additionally, we will discuss steps to soundproof a space, as well as headphones and monitor/speaker selection. This chapter also covers how to prepare for and perform an upgrade of Pro Tools 8 LE on an Apple Macintosh system, how to use the iLok key, and much more.

1.1 Introduction

Pro Tools (made by Digidesign) is a computer-based audio recording software program that runs on both Macintosh- and Windows-based operating systems. Pro Tools version 8.0 (the latest release) does not disappoint – its enhancements, new look and feel, and newly available tools make it one of the leading programs on the market today. In this chapter, we are going to start the process of planning, designing, and then building a DAW, which is the nervous system for your digital audio recording workflow and production process.

1.2 The Digital Audio Workstation

The DAW is where music is recorded, engineered, and produced. In the days before digital recording, engineers primarily recorded with analog components. While a DAW is primarily found within a digital domain, it does not mean that analog is no longer used. Although many argue the pros and cons of using either analog or digital, it is safe to say that a mix of both is usually the best solution when planning your DAW.

In this section, we will look at all of the items you will use to construct your DAW. We provide the blueprints to the most common designs and show you how to add external components such as analog gear. You should customize your DAW to your wants and needs. How you build yours is up to you.

A DAW is simply a computer system running Pro Tools (or other recording/ editing software) and connected to a hardware device that allows you to connect other devices such as electric instruments, microphones, Musical Instrument Digital Interface (MIDI) devices and instruments, and much more. Pro Tools 8 will only run with certain audio interfaces such as the Mbox2, which we will cover later. Although today's DAWs can be a simple laptop with a few peripherals, Figure 1.1 shows an example of an expanded DAW within a recording studio making use of a control surface, analog gear, and monitors within a treated listening space.

Figure 1.1 Viewing a DAW in a recording studio.

A key feature of DAWs is their flexibility, as you can build anything you need so that you have the ability to freely manipulate recorded sounds. Normally, the computer system needed to run the software requires a large amount of processing power, memory, and hard disk space. Because of this, extremely high-powered systems are used to run your basic DAW, so it's recommended that you purchase or use a computer that has the extra horsepower. Today, most PC- and Macintosh-based systems handle the task without issue. Although both platforms are available for use, most audio professionals prefer the Macintosh due to its proven time-tested stability. Pro Tools 8.0 LE currently runs on both Apple Macintosh OS X (Leopard) and Microsoft Windows Vista or XP.

You should refer to each platform's minimum requirements and exceed them based on the load you think you may carry on the system. ("Load" means the number of tracks, number of soft synths, number of plug-ins, and sample rate.) Session files are quite large, and hard disk space is eaten up quickly. When performing real-time operations on your DAW, you will see how quickly computer resources become drained.

Table 1.1 shows guidelines to help you plan your base system.

Table 1.1 Pro Tools 8 LE system requirements

Operating system	Windows Vista/XP – PC; Mac OS X – Leopard 10.5.5 or higher – Macintosh. Note that when using XP, the Home and Professional editions with Service Pack 3 have been tested and certified by Digidesign. When using Vista, Ultimate, and Business editions with Service Pack 1 have been tested and certified by Digidesign. You can use other versions, but they have not been certified as of the writing of this publication.
Processor	1 GHz or higher processor. Multiple CPU or multicore systems recommended for best performance.
Disk space	Ensure that you have enough disk space set aside for the operating system, Pro Tools 8 LE, and all the other software you will need to install, such as extras, plug-ins, and session files. You may also want to consider extra space for backups if you are doing them locally.
Memory	It is recommended that you use at least 1–2 GB of RAM and up to 4 GB or higher if you can afford it. Memory provides the best bang for your buck when using a computer system. The more memory installed, the better your performance.
Peripherals	You will need a keyboard, mouse, and monitor. It's recommended that you get a large widescreen monitor to handle the many windows and tools you will be using. You can also set up Vista or XP for multiple monitors, thus allowing you to offload tools to separate screens. You will need multiple video cards or a card with multiple interfaces to do this.

Note

These are just guidelines; as the more stress you put on your systems, the more power you will need. As noted above, Windows XP and Vista are both supported. If you choose to use Vista, it is recommended that you disable Windows Aero because it causes errors when you try to use lower hardware buffer sizes (explained later). Windows Aero is an alternate desktop interface you can configure for Vista that is processor-intensive.

For simplicity, we will simply refer to Windows Vista and all its offerings as "Vista" and Apple Macintosh (Mac) OS X Leopard as "Leopard." The examples within the book and Web site are also shown primarily using the Leopard system. Once you know what system you will use and how to plan it, you can install Pro Tools 8 LE and get right to work. Next, we need to lay out the plan for all the hardware you need to assemble to build the studio that is right for you before installing Pro Tools 8 LE.

1.3 Software and hardware planning

As mentioned earlier, you may only need a simple laptop, speakers or headphones, and a couple of Universal Serial Bus (USB) musical keyboards to complete your studio. Add in a pair of headphones and you have a DAW ready to go. Others require a high-powered computer system, an expanded control surface to do hands-on mixing and editing, and racks of outboard gear to bring in effects. Your needs may vary and, luckily, Pro Tools 8 comes in a few different configurations to meet those needs. In this section, we will look at what versions of Pro Tools 8 are currently available and which hardware devices you can use to start the assembly of your DAW.

Getting familiar with Pro Tools and how it can be used can be confusing if you are new to the program. A Digidesign audio interface is necessary as well as the correct version of Pro Tools to run with that specific device. Once you know which interfaces are available and which you need to use, selecting the version of Pro Tools will be easier. Since you have multiple interfaces to choose from, you will have to decide first what you want to use your DAW for and what types of options you need. When deploying Pro Tools 8, you have three options from which to choose. There are three major versions of Pro Tools available: Pro Tools HD, Pro Tools LE, and Pro Tools M-Powered. Table 1.2 gives a detailed breakdown of each offering.

Pro Tools HD systems provide far more processing power thus giving you superior performance. DAW's consume power so using HD-based systems allows you to use Accel interfaces that can offload Digital Signal Processing (DSP) to an external component. This frees your computer system to maintain the session and run Pro Tools. Because of this separation, you get more stability and increased performance. However, this doesn't mean you can't push Pro Tools LE and M-Powered systems just as hard! You are dealing with the same version of Pro Tools, just less functionality. With the release of Pro Tools 8 LE, you will find that it's been enhanced in ways (such as track count availability) that almost rivals the last version of HD! Although HD does have more horsepower, you can build your base computer system to withstand a large processing load and use Pro Tools 8 LE instead of HD, which gives you

Table 1.2 Pro Tools 8 software versions and offerings

Version	Offerings
Pro Tools 8|HD	Pro Tools|HD Accel Digital Audio Workstations, high-end I/O options, and control surfaces. Commonly found in professional and project studios. Used with specialized hardware to provide more performance and stability.
Pro Tools 8 LE	Affordable, powerful Pro Tools system that works with 003 and Mbox2 interfaces. Commonly found in home, mobile, professional, and project studios. This is the most widely used version of Pro Tools.
Pro Tools 8 M-Powered	Affordable, powerful Pro Tools system commonly found in home, mobile, and project studio systems that work with qualified M-Audio interfaces. Pro Tools 8 LE and M-Powered systems are marketed side by side and offer the same functionality.

more planning and design flexibility. No matter the version, anyone working with Pro Tools in a production environment will tell you first hand; "power" is what you need to get things done. Without processing power, you may experience many problems while working and pushing your systems to their limits.

Now that you understand what each version of Pro Tools 8 gives you, the next step is to choose an appropriate audio interface.

1.4 Understanding Pro Tools hardware offerings

Digidesign offers many options when it comes to choosing an audio interface. Most options are based on how many and what types of external connections you will make with it, as well as the quality of electronics contained within it. For example, if you need to install a recording studio and are considering future growth, you can purchase the 003 family of products. If you need the smallest hardware footprint available, the Mbox2 Micro is the size of a stick of gum. Table 1.3 outlines the current audio interface offerings.

As you can see, there are many offerings, and you can select the right hardware based on price and need. For example, if you find you have a need for only one XLR-based microphone preamp, then the Mbox2 Mini is for you. Note, however, that the hardware you choose unfortunately is not modular and therefore not scalable. You cannot add more preamps, although you can change out the ones currently in the interface. That's why it's important to design your DAW correctly, as you may find you outgrow your initial purchase quickly depending

Table 1.3 Digidesign audio interfaces	
Mbox family	
Mbox2 Micro	This device does not provide an audio input. It does provide a 1/8-inch stereo output for headphone or speaker use and a volume wheel on the side. Used for on-the-go mixing and editing of Pro Tools sessions.
Mbox2 Mini	This device offers two 1/4" or 1 XLR input for recording; Universal Serial Bus (USB)-based. XLR-type connections allow for high-quality electrical connections between microphones and audio equipment.
Mbox2	This device offers 2 XLR inputs and more functionality on the front panel of the unit; USB-based. USB is a commonly used PC-based connector type that allows for fast connections between the PC and peripherals.
Mbox2 Pro	This device offers FireWire and a MIDI interface. FireWire is a significant upgrade to USB and allows for much higher transfer speeds.
003 family	
003 Rack	Offers 4 XLR inputs with high-end A/D converters, excellent preamps (with front panel gain controls), eight line inputs and outputs, a Word Clock, a MIDI interface, S/PDIF, and an ADAT light pipe connection. S/PDIF and ADAT connections allow for high-speed digital transfer of data between devices.
003 Factory	Same as the 003 Rack but offers more software plug-in options.

on what you are looking to accomplish. Figure 1.2 shows the Mbox2 Micro, the newest and smallest audio interface available from Digidesign today.

Figure 1.2 Mbox2 Micro.

The Mbox2 Micro is easy to use and can connect to a USB-capable computer quickly and easily. This device lets you take a session out on the road to mix and edit it. You will not be able to record directly with this unit, but you can load up and use Pro Tools while on the go. As you may find in your own travels, many working producers and engineers rely on being mobile because there are so many different places in which you can work and collaborate globally. A previous restriction to using Pro Tools was that when you wanted to go on the road, the Mbox2 was the smallest device you could take, and unsurprisingly, it is unwieldy to power up and use on the go. Now, with the Micro, you can pop in the device and get to work immediately with the smallest profile imaginable.

Digidesign also released the Mbox2 Mini, which is what you would look to purchase if you needed the ability to record. The Mbox2 Mini is an audio interface, which provides easy on-the-go recording, editing, and mixing. Figure 1.3 shows the Mbox2 Mini.

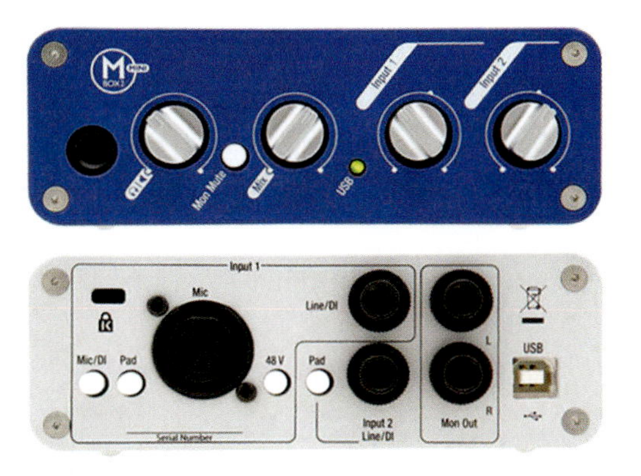

Figure 1.3 Viewing the Mbox2 Mini.

As you can see from the rear of the Mini, you have one XLR connector to plug a microphone into, as well as two Line/Direct Injection (DI) inputs for a 1/4-inch balanced or unbalanced cable. You can plug a keyboard, drum machine, electric guitar, or electric bass into the DI inputs using a cable under about 15 feet. If those instruments are farther from the interface, it's better to use a separate Direct Box (DI). It converts the instrument's unbalanced high-impedance signal in a phone plug to a low-impedance balanced signal in an XLR connector. That prevents hum pickup and high-frequency loss in long cables. In live sound situations, using direct boxes instead of microphones reduces leakage from other instruments if acoustical isolation is not possible.

In addition to the DI inputs and the XLR-based connection, the Mini also has left and right outs. What the Mini does not include is 48V phantom power for condenser microphones and a MIDI interface. The Mbox2, Mbox2 Pro, and 003 family do include those features. Figure 1.4 shows the Mbox2, the original upgrade from the first Mbox.

When the Mbox2 came out, it offered a sleeker design and better circuitry among other upgrades from the original Mbox. A quick glance to the rear of the unit shows S/PDIF connections and MIDI I/O (input and output) though a new integrated MIDI interface. As you will find, S/PDIF lets you transfer high-quality digital audio in real time. Word Clock I/O might be needed to keep a large digital system in sync.

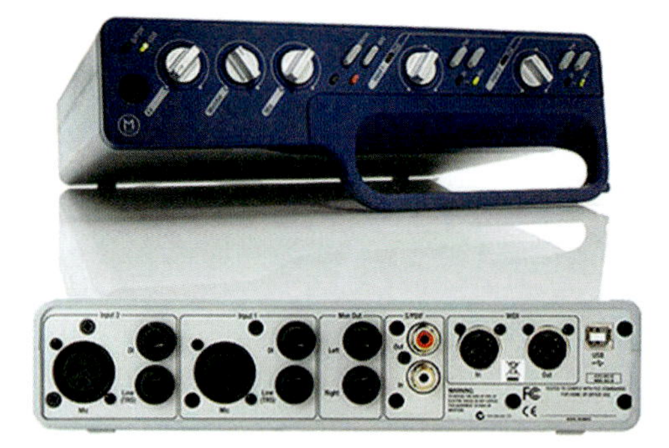

Figure 1.4 Mbox2.

Compared to the Mini, the Mbox2 offers another XLR-microphone connection and another DI connection. This opens up more flexibility in recording, as you can record more than one track at a time. This unit is great to take with you on the road to capture outstanding sound when used with stereo microphones. If you need a little more power, you could choose the Mbox2 Pro, the scaled-up version of the Mbox2, which offers more options and flexibility with two headphone jacks for simultaneous playback on each, connections for an external Word Clock and a connector for an external power source. Figure 1.5 shows the Mbox2 Pro.

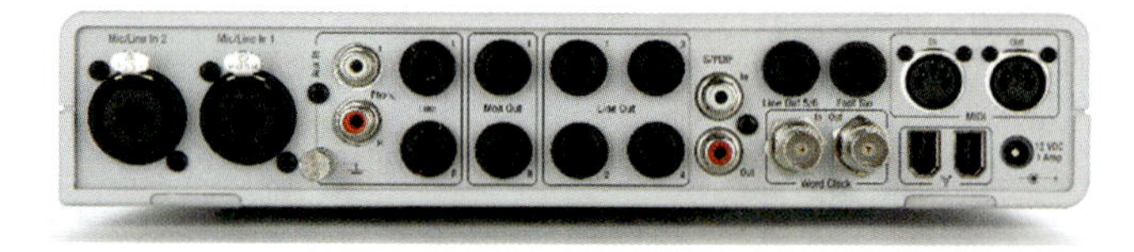

Figure 1.5 Mbox2 Pro.

With the Mbox2 Pro, you now have the ability to connect to your computer via FireWire. Each offering up until now allowed you to use a USB 1.1 connection. With FireWire (400 or 800), you can transfer data at increased speeds of up to 400 or 800 Mbps.

That sums up the offerings in the Mbox2 family. Many engineers do not need more than this to do what they want to do, and many project studios today run on the Mbox2 or Mbox2 Pro. If your needs outgrow these two devices, then your next step is to purchase a device in the 003 family of products, the upgrade to the original Digi001 and Digi002 systems.

The 003 line is nothing more than a rackable version of expanded Mbox technology. The components such as preamps and connecting circuitry are obviously of better quality, but the main draw for this unit is the amount of connections you can add to it. The 003 line is the first step into Digidesign modular designs. For example, with the 003, you have an optical connection allowing external preamps and other devices to be chained and used at high speed. As with any upgrade path, the 003 line offers better equipment and more options and flexibility than the Mbox2 family of products. Figure 1.6 shows the 003 Rack.

Figure 1.6 003 Rack.

From the front and rear of the 003 Rack, you can see the many additional inputs and outputs and expanded options and features not found in the Mbox2 family.

Note

Any sessions created with any version of Pro Tools will open on any version of Digidesign hardware. The only caveat is that HD system sessions will not open on Pro Tools LE or M-Powered systems and be completely functional.

Using a Word Clock

A Word Clock is used to sync-up your digital devices that need to transfer digital audio in real time. All devices in the digital chain must use a synchronized sample rate that comes from the master clock. The Word Clock signal is used by many devices such as the Mbox2, the 003, external components, such as mastering decks, disc players, and much more. This clock source works with many interface types such as Alesis Digital Audio Tape (ADAT), S/PDIF, and AES/EBU to name a few. The Digidesign series of audio interfaces supply a master clock option configurable in the Hardware Setup dialog box found in the Setup menu. To change this option, go to Setup → Hardware → Hardware Setup Dialog box → Clock Source: field.

When configuring your hardware, you should consider that all your devices need one source for clocking for synchronization. If you have external devices connected to your interface through the ADAT Optical or S/PDIF inputs, the master clock should be derived from the external device. The configuration can be found within Pro Tools 8 LE by going to Setup menu → Hardware and then setting the clock source to ADAT or S/PDIF depending on the type of connection used. If you do not set this correctly, you may suffer from jitter or other issues that will wind up on the recorded work. Jitter is a variation in the timed signal that, if severe enough, could create errors in your recorded work.

Now that you have learned about the many software and hardware offerings you can choose from, let's finish our DAW by adding other devices, peripherals, and components. Once you have all of your devices selected, you can build your DAW and install Pro Tools.

1.5 Understanding Pro Tools-based control surfaces

Digidesign offers a host of components called "control surfaces," which allow you to manipulate and use Pro Tools without having to be confined to a keyboard and mouse. Although you do not need a control surface to use Pro Tools, it definitely helps when you are working on larger projects. Current offerings include the Command|8, C|24, and ICON control surfaces. Table 1.4 covers the major differences between each.

Table 1.4 Viewing Digidesign control surface offerings

Command 8 (Command|8)	The Command|8 is Digidesign's entry-level offering providing mixing and editing options with a control surface and your keyboard. It allows for automation among other features.
Control 24 (C|24)	The C|24 (previously known as the Control 24) is Digidesign's mid-level offering for those who want complete control over their sessions while using a control surface. The C|24 offers its own set of preamplifiers that rival those found in the 003 and offers the user more complete control compared with using just the keyboard.
ICON	ICON D-Control and D-Command control surfaces are Digidesign integrated consoles that provide the ultimate in control and expanded functionality, high-end circuitry, preamps, and much more.

Note

Venue systems are used for live performance and are very similar to the ICON family. You can find more information on Venue-based systems from Digidesign at Digidesign.com.

The Command|8 is also offered with the 003 as a complete package for those looking to save space and/or build a smaller scale project studio. Figure 1.7 shows the 003-based Command|8 Control Surface. As you can see, this unit would fit nicely on top of a desk.

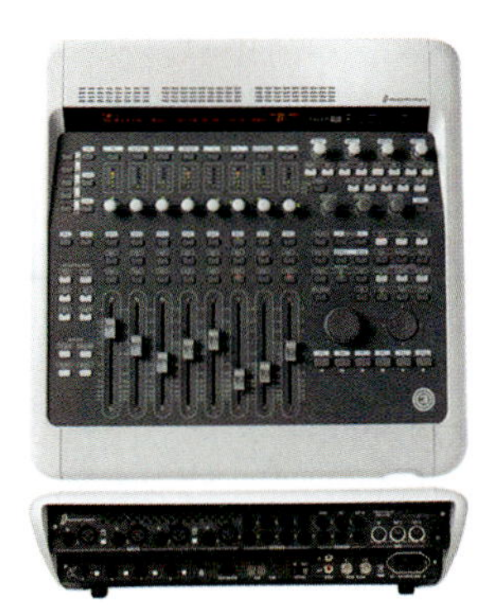

Figure 1.7 Command|8 Control Surface.

The C|24 (previously known as the Control 24) is Digidesign's mid-level offering for those who want complete control over their sessions while using a control surface. The C|24 offers its own set of preamplifiers that rival those found in the 003 and offers the user a more expanded and complete control as compared to a keyboard. Although working with both can provide many benefits, you will find that this control surface gives you the most flexibility for its price point. Figure 1.8 shows the C|24 Control Surface. You can use this console with a Pro Tools HD- or Pro Tools LE-based system.

Figure 1.8 C|24 Control Surface.

The one caveat to using this type of surface is that you will need to either find a large desk to place it on or purchase a desk that has the "cut-out" for mounting and installing the C|24. Most audio/producer desks come with the proper sizes already in place, as well as for empty spots for outboard gear, other components, speakers, and your computer equipment.

Finally, if you find the need to interconnect offices via Satellite links and control multiple studios at once, transfer data between them, and have complete control, flexibility, and power located at your fingertips, the ICON D-Control and D-Command control surfaces are for you. The ICON integrated console environment, featuring the D-Control and D-Command "worksurfaces," includes Pro Tools|HD Accel as the core DSP engine and modular Pro Tools|HD audio interfaces. The ICON series of control surfaces are integrated consoles that provide the ultimate in control and expanded functionality. If your budget allows for it, then this is the control surface that provides the most power and functionality. However, it must be used with a Pro Tools HD-based system. Figure 1.9 shows an ICON control surface.

Figure 1.9 ICON control surface.

Once you have selected the control surface that fits your needs and your budget, the next step is to select the proper cables and get everything hooked up correctly.

The importance of ergonomics

When working as an audio engineer or a producer, you will find yourself to be quite the contortionist. It's very important to work correctly and comfortably. If you are a working engineer, you are pushing a lot of buttons, moving the mouse, hitting the keyboard, and fiddling with knobs, instruments, and cables. You will be moving constantly, and the more comfortable you are, the less chance of injury and the easier your life will be. You should sit in a comfortable chair and practice good posture – setting up your monitors, screens, and equipment correctly is the key.

For example, a chair with lumbar support is recommended. Also, the height in which you sit, how far your monitor is away from view, or if it's positioned to the left or right can all cause stress in your body. Typing can also cause repetitive stress disorder such as carpel tunnel. If you position yourself in such a way that you are not comfortable, you could wind up with pinched nerves, a stiff back and neck, and other problems.

1.6 Cable management

You must know about cabling to work with Pro Tools. External hard drives, Digidesign peripherals, MIDI equipment, and many other devices are all connected by cables. There are many types, but they are easily identified and understood. In this section, we cover the most fundamental cable types you will need to know about to get Pro Tools 8 LE ready for use.

First, you must know where all your components will be to get cables of the correct length as well as type. You must also know that some cables are more resistant to problems than others, some use a shield while others don't, and there are many different connectors from which to choose. When designing a DAW, the first set of cables you need will connect your computer to either the Mbox2 or 003 family of products. As mentioned earlier, you will find that the Mbox2 Pro connects only via FireWire, whereas the Mbox2 connects only via USB. Table 1.5 outlines the most critical types of cables.

Table 1.5 Cable types

USB and FireWire	Two computer-based connectivity options for your Digidesign hardware. FireWire connections should be used whenever possible to provide for the highest transfer rate. This will provide you with a smoother session, reducing the possibility of hang-ups that may occur from using USB, which is slower.
XLR and 1/4″ cables (balanced and unbalanced)	These cables provide an analog connection from an instrument (or microphone to a DAW audio interface), where their signals are converted to digital for processing. Balanced cables provide for longer runs without hum pickup or high-frequency loss, which occurs in unbalanced cables over long distances.
Optical (Lightpipe)	Used on the 003 and other third-party components for expansion, such as additional microphone preamps.
S/PDIF	Supplies a direct digital connection to devices such as a digital mastering deck.
Ethernet	Control surfaces such as the C|24 use an Ethernet interface to connect to the 003.

It's important to use the correct cable because using balanced or unbalanced cables can dramatically affect your sound. With a balanced cable (usually a 1/4″ TRS or 3-pin XLR cable), a signal is carried by two inner conductors and the shield is not part of the signal path. Interference picked up is induced to both conductors and canceled. Unbalanced cables (usually a 1/4″ TS or RCA cable) use a center conductor as the hot wire and the shield as the cold and are prone to interference in length in excess of 10 feet. Use balanced connections whenever possible and especially when you want to avoid interference in long cable runs. Balanced XLR cables are used to connect the most professional audio equipment.

It's very important to know how ADAT Optical (also known as Lightpipe) and S/PDIF inputs work, as often times your expansion options require intimate knowledge of their use. Using the ADAT Optical connection, you can connect any ADAT-compatible device to up to eight more inputs, such as the DigiMAX LT Eight-Channel Microphone-Preamplifier with ADAT Lightpipe Output (http://www.presonus.com/). The S/PDIF jack on the back of the 003 offers an additional two inputs. When you add up all the additional inputs, you will have 18 inputs in total.

Interconnecting devices between different parts of your studio (such as a vocal booth and control room, for example) requires the use of long cable runs connected via patch panels or mic snakes. Figure 1.10 shows a 1/4"-connection-based patch panel.

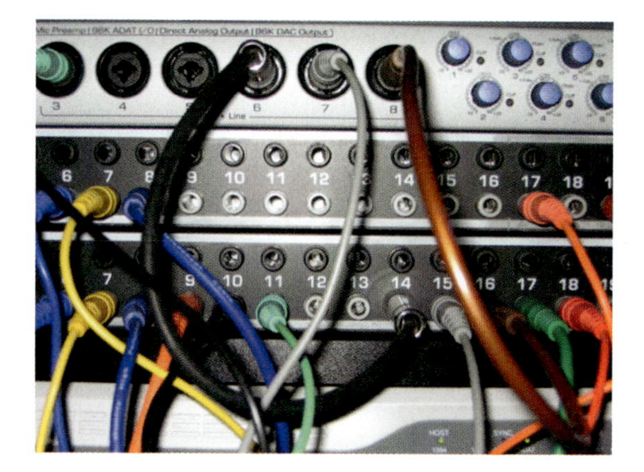

Figure 1.10 Patch panel (1/4").

You can now connect microphones, speakers, instruments, and much more by simply patching them in. Make sure that you use high-quality connections to ensure the integrity of your digital signal. Longer runs may produce a failed or corrupted signal, and any power source disruptions such as electromagnetic interference (EMI) can also produce problems for your production. Do not run cabling over any fluorescent lighting or near or over power sources, and use only high-quality cabling whenever possible. Do not exceed any set limitation on distance, especially when using technologies such as Ethernet.

1.7 MIDI hardware

As we will learn in Chapter 3, composing often requires the use of the MIDI standard. MIDI is a technical name for a standard that is used to send signals back and forth between interconnected MIDI devices. A more thorough discussion of MIDI is in Chapter 3, but for now, if your goal is to score and produce music, or write a piece of music on a piano (for example), you may want to set up a MIDI interface at this point.

For those who use a piano to write music, MIDI hardware such as the M-Audio Keystation series provides additional functionality for creating all types of music. It includes features such as a pitch bend wheel to change the pitch of anything you play. Figure 1.11 shows the M-Audio Keystation.

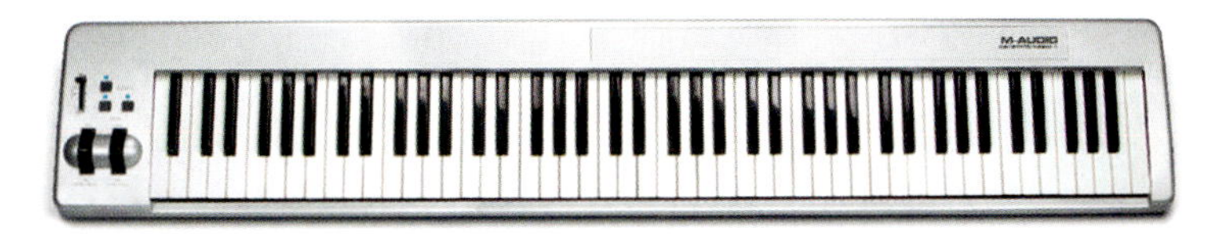

Figure 1.11 M-Audio Keystation.

There are multiple offerings in the Keystation series that offer different features to meet all types of budgets. There are also smaller MIDI interfaces offered for mobile travel if you find a full-sized keyboard a challenge to travel with.

1.8 Microphones

The next component you will need is a microphone ("mic"). There are many types of and uses for microphones and different microphones for different projects. What you intend on capturing should be reflected in the actual recording and getting the right microphone for the job will ensure that. The microphone you choose for your recording project will have a large effect on the quality of the recording you capture and the overall sound of the instrument you may be trying to capture. Select your mics carefully and choose what works best for you and the source you are recording. In its simplest definition, the microphone (Fig. 1.12) can be defined as an electronic ear. A more complex definition of a microphone is that it is a transducer (a device that changes energy from one form to another) that converts sound information from patterns of air pressure into electric impulses.

A variety of mechanical techniques can be used in building microphones. The two most commonly encountered in recording studios are the magneto-dynamic and variable capacitance designs. You would commonly see these two called dynamic and condenser for short. The type of mic most commonly used for capturing a live performance is the dynamic microphone. You

Figure 1.12 A prepped microphone in a vocal booth.

can find a number of dynamic mics in most professional and project studios worldwide. A dynamic microphone is most useful when the sound source is close and reasonably loud. It may be your best choice if you want to record sounds that are predominantly bass or mid-range. When it comes to very high-frequency sounds such as cymbals, bells, or the upper harmonics of the acoustic guitar or piano, condenser mics are better suited, as they are more sensitive to higher frequencies. For instruments with lots of top-end detail or which output quieter sounds, condenser microphones (also known as capacitor microphones) are unsurpassed. Studio vocals are generally recorded with condenser mics that may cost thousands of dollars but are worth it when attempting to obtain the highest quality. Condenser microphones are also known for their good transient response and detail and are useful for capturing instruments such as acoustic guitar. Ribbon mics are different from condenser mics in that they are less sensitive to high-frequencies, and they are most effective when recording electric guitar amps, horns, acoustic bass, and sibilant vocals.

When selecting a microphone, you may also want to consider its frequency response. Frequency response is the range of frequencies that the mic can reproduce within a certain tolerance, such as ±3 dB. All microphones exhibit different frequency responses. These differences in response can make certain microphones more suitable than others for specific purposes. For example, a microphone with a frequency response that favors high-end frequencies will sound "bright" and could be useful on an instrument such as cymbals.

It is also important to note that certain microphones can distort from high sound pressure level (SPL) if it exceeds the tested limits of the mic. SPL is the measurement of the amplitude (loudness) of sound, measured in decibels (dB SPL) relative to the threshold of hearing (0 dB SPL). For example, certain condenser microphones might distort with loud sources such as drums or electric guitar.

When recording, you will need to make sure that you have the proper microphone type and placement before starting the actual recording. It may seem like a lot of work to do before you start the recording session, but rest assured – every step you take while selecting and placing your microphones will help in capturing the best possible recording from the artist(s). Placement of microphones and then setting their levels properly are essential to capturing workable material. Maintaining the appropriate recording levels will preserve the dynamics, lessen the chance for digital clipping, and ultimately result in a better quality recording. Constant monitoring of the work in progress will keep you ahead of the process and ultimately save you a lot of time and headaches. Placement is easier when you know what you are trying to capture and what configuration will capture what you intend to reproduce. For just the drums alone, you could use from one to over half a dozen microphones, so proper preparation is critical.

In Chapter 4, we will discuss some common ways to place microphones for use when conducting a recording session.

1.9 Triggers and drum processors

Drums (as well as many precision instruments) can be sampled. Sampling a drum sound is nothing more than recording the hit of the drum and then replacing the sound from a library of sounds you can access in a DAW such as Pro Tools. These files are usually of high quality and sometimes of better quality than the actual sounds you record live. A trigger system is made up of a brain and trigger units – also known simply as "triggers." The trigger is a type of transducer, which reacts to pressure changes caused by the strike of an acoustic drum or electronic drum pad. This strike is then sent to the brain where it can be further processed. You can assign just about any sound imaginable in the brain, depending on the unit you purchase.

Many people today use triggering systems in their DAWs, but many recording studios prefer the sound of a live drum kit, as triggered drum samples are not able to recreate room sound better than an acoustically designed live room. Recording the drums' live sound allows engineers to use many techniques to create unique sounds that are not available with triggered drum samples. The choice of which route to take is ultimately yours. Figure 1.13 shows a set of triggers and a sound module.

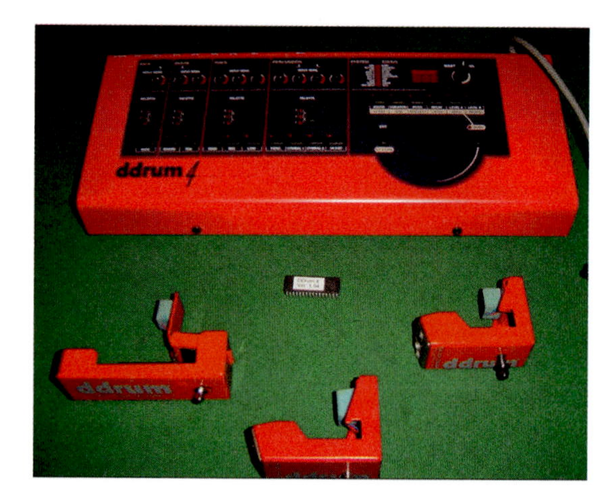

Figure 1.13 Triggering system.

1.10 Using outboard gear

All studios expand their sound production options by using outboard gear. What is this? Outboard gear is nothing more than effects processors, components, and peripherals that you bring into your recording, editing, mixing, and mastering processes. One example of outboard gear is a microphone preamplifier (or preamp for short), a module that you can use to amplify and perhaps modify a microphone signal. Figure 1.14 shows a Vintech X731 2-channel microphone preamp. It might be an improvement over the mic preamp in your 003 or Mbox2 audio interface. Simply setting up your audio interface's inputs and outputs (I/O) correctly will allow you to add any of these thousands of devices, including the preamp shown in Fig. 1.14 and other devices such as compressors, limiters, noise gates, and effects units, to your DAW. The next chapter will cover in more depth the use of outboard gear with Pro Tools, including signal routing and I/O.

Figure 1.14 Microphone preamp.

1.11 Headphones and monitors

No DAW is complete without a pair of stereo headphones. Headphones vary in quality of sound and comfort, so it's important to read reviews, try a pair if you can, and always remember to consider comfort in your purchase for those long sessions. For mixing and final tweaking, obviously comfort takes a back seat to sound quality, but ideally you should find a pair that meets both needs.

When connecting monitoring devices to your Digidesign hardware, you have the option of connecting monitors (speakers), headphones, or both simultaneously. Headphones are used when you want to record overdubs or to concentrate on the playback without bothering anyone around you.

When connecting your speakers to the Mbox2, you may use the "line out" jack to feed the external amplifier for your monitors. If your amplifier does not have its own volume control, you may instead use the rear headphone jack to feed your monitors. Simply use an adapter to split left and right into separate channels and then connect to your amplifier. On the 003, use the "mon out" jack to feed your external amplifier. Use the "monitor level" knob located on the front panel to control volume. The "alt main" jack may be used to bypass the volume control of the 003. To connect headphones, for the Mbox2, use either the headphone jack on the front or the rear to feed your headphones, and for the 003, use the headphone jack located on the front panel.

In an acoustically treated room, flush mounting your speakers to the wall, also known as Soffit mounting, offers many important advantages, such as eliminating back-wall reflections, minimizing diffraction caused by the speaker cabinet edges, and boosting speaker output. Flush mounting also improves transient response and speaker imaging. Speaker cabinets should be mounted symmetrically, approximately ear height, at a 60-degree angle from the mix position. When setting the speakers at a 60-degree angle, the distance from the mix position to the speakers must be equal to the distance between the speakers. The front of the speaker must be exactly flush with the front wall, with a maximum gap of 3 mm between the edges of the speaker cabinet and the opening in the wall. It's important to note that the speaker must not make direct contact with the front wall. Speakers may be mounted on stands inside the wall. Both the speaker stands and the wall they are mounted in should have sufficient weight, for example, a speaker stand made of concrete or sand is appropriate. The speaker should then be floated on a rubber sheet to further isolate the room from vibrations caused by the speaker cabinet.

Tip

> Flush mounting forces speakers to radiate in a semicircular pattern, eliminating rear reflections and focusing speaker response.

1.12 Treatment and soundproofing

For those working in a production studio, you likely already have soundproofing and acoustic treatment in place. Soundproofing is isolating a room from outside noises. Acoustic treatment is the placement of acoustic absorbers on the room surfaces to prevent excessive sound reflections off the surfaces. These measures help produce the best environment possible for your recording, editing, and mixing experience. Treating the room for noise reduction and isolation, along with placing the microphones effectively, is very important to capturing the intended sound. In a control room, room reflections cause cancellation at certain frequencies, thus blurring the accuracy of the material you are listening to. Eliminating initial room reflections provides a clearer image of the sound source and therefore high-quality sound on all types of playback systems. When you have an isolated environment free from exterior noise, you will be ready to place and cable all the microphones to your Pro Tools system and be confident that you will capture only what you intend to capture.

Whether building a project studio or a high-end recording studio, all effort should be made to do the job right. You can reduce the materials cost by carefully selecting where you shop and carefully selecting cheaper alternatives, but for the most part, your material list will be standard. An absolute must when applying isolation treatment is to achieve an airtight seal. Designing and erecting a room without parallel walls reduces flutter echoes – sound reflections between opposite walls. Make sure that you use materials that are acoustically tested and made specifically for sound proofing projects and consider safety every step of the way. Fire safety is paramount so be sure not to block entrances or create dangerous situations when treating a room or space.

Excess low frequencies or standing waves in a room can obscure focus and smear timing information, covering up the details of your recording. One of the most commonly used solutions is a bass trap. Simple to make and highly recommended, a bass trap is sound-absorbing material used to diminish low-frequency standing waves. Most bass traps consist of dense material. A simple bass trap can be created without much difficulty. Use a 2′ × 4′ sheet of rigid fiberglass such as Owens Corning 703 in 4″ thickness covered in fabric spanning the corners of the room. In the materials list, you can substitute Rockwool in place of rigid fiberglass to keep costs down.

When planning your studio, consider size and ergonomics, as the area also needs to be comfortable for you and your guests. Consider airflow, heating, and cooling solutions. If you are going to build a simple sound booth, you need to take the same level of care you would for building a very large recording studio. Any external noise will wind up in the recording, so it's essential to use double-wall construction or concrete blocks to isolate the studio from outside sounds. Even for a home-based studio project, you can apply the same concepts as would be used in a project studio by simply scaling down the size and cost. For example, use thick comforters, convoluted mattress foam, or sleeping bags for acoustic absorption at mid-to-high frequencies.

Note

Auralex Acoustics (http://www.auralex.com/) is one of the industry leaders in acoustical treatment products, including acoustical absorbers, diffusers, sound barriers, construction materials, isolation platforms, and complete room treatment systems. You can learn a lot from their Web site and find options within your budget to treat any space.

So, now that we have the DAW ready to go, let's install Pro Tools!

1.13 Getting Pro Tools 8 LE

Pro Tools 8 LE can be purchased at an audio equipment store or online. You can also choose to download it directly from Digidesign through Digidelivery. After purchase, you must register the copy online and authorize it for use. We will cover these steps in the next section.

Digidelivery

Those of you who send work to and from Digidesign directly, Digidelivery is Digidesign's method to transfer files and data. This is also integrated directly into Pro Tools where a menu option allows you to connect direct. Figure 1.15 shows the use of Digidelivery to download Pro Tools 8 LE for installation.

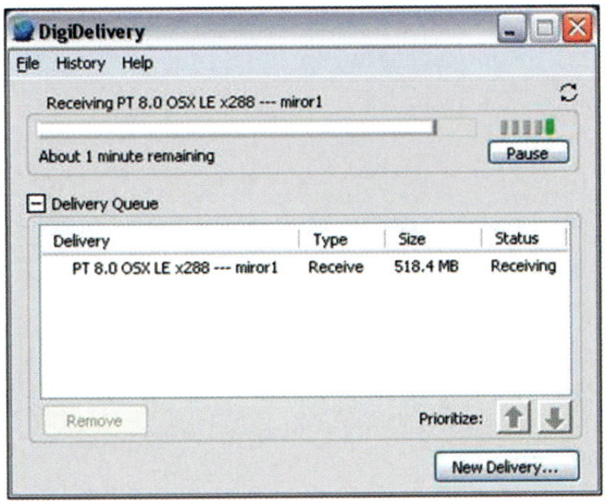

DigiDelivery

File History Help

Receiving PT 8.0 OSX LE x288 --- miror1

About 1 minute remaining

Pause

Delivery Queue

Delivery	Type	Size	Status
PT 8.0 OSX LE x288 --- miror1	Receive	518.4 MB	Receiving

Remove

Prioritize:

New Delivery...

Figure 1.15 Viewing a digital download of Pro Tools 8 LE.

Once you have either purchased or downloaded a copy of Pro Tools 8, you will need to prepare it for installation.

1.14 Using iLok

To install Pro Tools 8 LE, you need to use an iLok key. The iLok key (http://www.iLok.com) is a USB hardware device that authorizes the use of software and prevents the illegal copying of software. It also allows the transfer of licenses between accounts for a small fee. The iLok key (Fig. 1.16) also allows you to use your plug-ins anywhere. So, if you rent or use a studio that is not your own, you have access to your software as long as you have your iLok key.

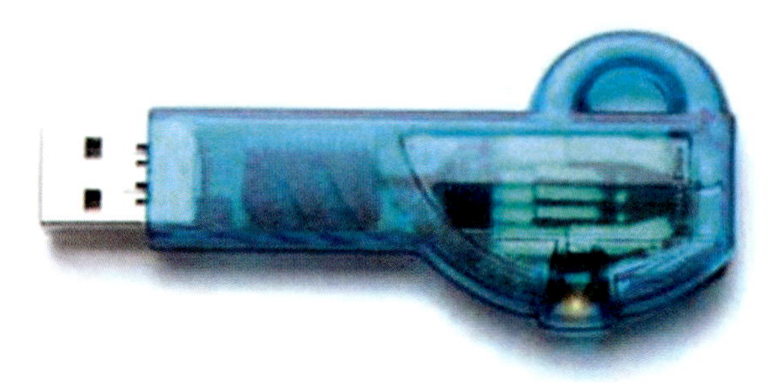

Figure 1.16 iLok key.

You can create and log into an online iLok account to make adjustments, transfer licenses, and check on the status of new licenses. Figure 1.17 shows the online iLok account and the interface where you can download new licenses and manage all your purchased software.

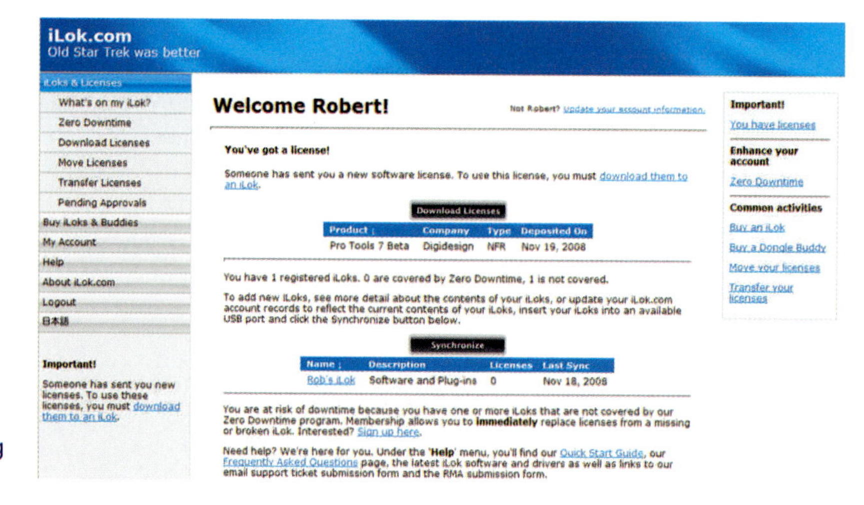

Figure 1.17 Managing an iLok account.

Once you have purchased Pro Tools 8 LE and prepped your iLok for use, you can get Apple OS X or Windows XP/Vista setup for your installation.

1.15 Upgrade Apple OS X

Anyone running an operating system older than Leopard (OS X 10.5.4) will *not* be able to install Pro Tools 8 LE. It will not install on anything other than Leopard. If you do not have Leopard, you will have to purchase and install it to use Pro Tools 8 LE.

1.16 Prepping the system before installation

Before you install Pro Tools 8, you will need to spend a few minutes tweaking Leopard. For example, there are services you should turn off, and other items you should change before even starting the process of installing Pro Tools 8 LE.

When preparing your computer to run Pro Tools, you should always keep unneeded applications off of it. As mentioned earlier, these will only take up needed resources. Pro Tools is a resource-intensive application, and it relies heavily on the system's CPU for its processing power. In addition to unneeded applications, there are services (or daemons) that run in the background. These services also take up RAM and may reduce the performance of Pro Tools. To make most of the changes to your Leopard system, simply open up System Preferences (found in the Apple menu) as seen in Fig. 1.18. You can also check system resources within Pro Tools itself. You can find the System

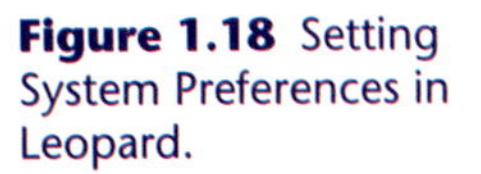

Figure 1.18 Setting System Preferences in Leopard.

Usage window by selecting System Usage from the Window drop-down menu when first opening Pro Tools.

On the Mac, you will want to disable the Spotlight indexing service and the Dashboard Shortcut, both of which can cause interruptions during your recording session. Spotlight is used to create an index of data on your computer useful for searches and will begin indexing any volume as soon as it is mounted. There are some exceptions, however, such as when using CDs and DVDs. The Spotlight feature runs in the background and affects system performance. To disable it, go to the Apple menu and select System Preferences. You can find Spotlight in System Preferences. Open Spotlight and click on Privacy. When you want to disable the indexing of a drive, simply drag the drive's icon into the Spotlight privacy list. Spotlight will no longer index this drive. You can see an example of this in Fig. 1.19.

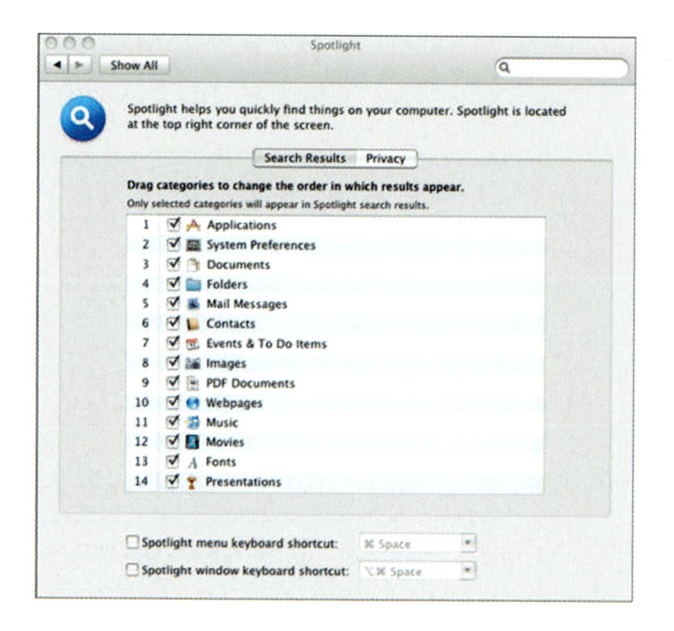

Figure 1.19 Adjusting Spotlight.

Once Spotlight is turned off, you should then disable the Dashboard key commands, which contains some keyboard commands that overlap those used in Pro Tools and must be disabled. For example, the Dashboard command normally assigned to the F12 key is also used by Pro Tools to start recording. To disable the Dashboard key commands, go to System Preferences and choose Dashboard and Exposé. You can disable each Dashboard keyboard command by setting it to Null (Fig. 1.20).

Now that your system is ready, let's discuss whether to use an external drive or the computer's internal drive to record your sessions. One will provide better performance than the other.

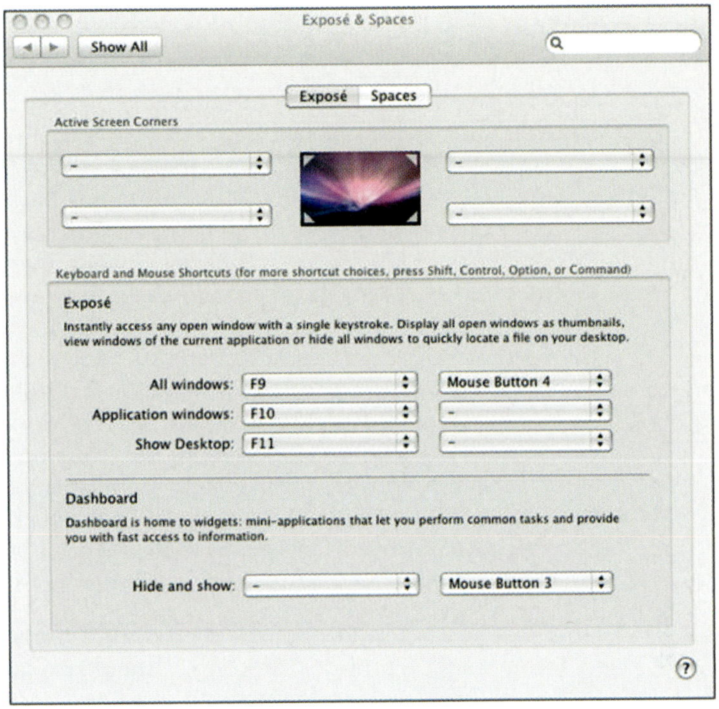

Figure 1.20 Adjusting the Dashboard.

1.17 Configuring an external drive

The last step you should take will make your Pro Tools experience better. It's a proven fact that running your sessions off a FireWire-based external hard drive, or a 7200RPM internal Enhanced Integrated Drive Electronics (EIDE) or Serial ATA (Serial Advanced Technology Attachment), or commonly shortened to SATA, drive that is separate from your operating system drive, will speed up your Pro Tools experience. Although Pro Tools will let you record to your system drive, this is generally not recommended. You should record to system drives only when necessary – for example, if your computer system has just one hard drive or your other hard drives are completely out of space. Also, to use your drive with multiple file systems (for example, Windows and Macintosh), it's necessary to format your drive a certain way (explained later) so that you can use the drive between systems.

When selecting drives and preparing them for your Pro Tools projects, get a drive that is officially certified by Digidesign and offers high performance and capacity. Supported drives and other helpful information can be found within the Digidesign hardware compatibility matrix found at http://Digidesign.com/compato. Digidesign tests drives and certifies that the models listed will work with Pro Tools.

External FireWire drives are a popular choice for storing Pro Tools projects. FireWire, also known as IEEE 1394, is a protocol for transferring data to and from devices at high speed. Originally invented by Apple, FireWire comes in two speeds: 400 and 800 Mbps. FireWire hard drives are portable, easy to connect, and fairly inexpensive. To record projects in sample rates up to 48 kHz, a single hard drive is sufficient, while a second drive is necessary to record projects at 96 kHz. A second drive is often also useful for backing up important data. Digidesign recommends FireWire 400 drives that contain the Oxford 911 chipset and FireWire 800 drives that contain the Oxford 912 chipset. Drives should also have a speed of 7200 RPM or better. Multiple drives may be used, but FireWire 400 and 800 drives should not be combined. When using FireWire 400, up to eight daisy-chained drives are supported, and when using FireWire 800, up to four daisy-chained drives are supported. To install external FireWire drives, be sure that you have the appropriate cable and that a FireWire IEEE 1394 port exists on your computer. If you are using a PCI FireWire card, check to make sure it is compatible with Pro Tools.

Now that you have selected the type of drive you will use, prepare it for use by formatting it. For Mac OS X, drives should be formatted using the Mac OS Extended (Journaled) format. Use the Disk Utility (found in the Utilities folder in Mac OS X) to enable journaling. You can get to the Disk Utility via the Go menu, by selecting Utilities and then scrolling down to Disk Utility. The Disk Utility window is seen in Fig. 1.21.

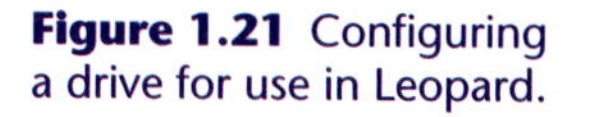

Figure 1.21 Configuring a drive for use in Leopard.

Simply click on the drive and select Erase. Note that your previous data will be removed, although Leopard gives you options to recover it if needed. Once your drive is formatted, it is ready for use with your computer. For best performance, the hard drives should be erased and formatted annually. Most hard disk manufacturers provide instructions on the installation and use of the drive, and they also may include software backup and disk maintenance utilities that should be used to keep data backed up. In Chapter 7, we will cover how to safeguard your data in multiple ways.

No matter what type of drive you choose to use, make sure you have enough space to store the material you are recording. Lack of disk space may cause you to experience poor performance and can cause your system to run slowly or in the worst case scenario, not at all. In Pro Tools, the Disk Usage window can be displayed by going to the Window menu and selecting Disk Space from the drop-down menu. You can display the Disk Usage window so that as you record and save your sessions, you can accurately monitor your hard drive resources. You will want to check this prior to recording to ensure that you have the resources needed for the entire recording session. The Disk Usage will tell you how much space you have remaining in percentages as well as minutes.

Time Machine

It is recommended that you turn off Time Machine when using Leopard. If you do not need it, then do not use it. It's recommended that any system you use for Digital Audio be designated as a computer only used for that purpose. This way you can not only turn these services off but also forgo the use of antivirus software that basically robs your entire system of its power to keep it "safe." If you go online infrequently and only for doing business, it's likely your system will not be the victim of an Internet-based attack. Figure 1.22 shows Time Machine, which can be used to backup your system if needed, but again, it's recommended that you do not use this with your Pro Tools session drive.

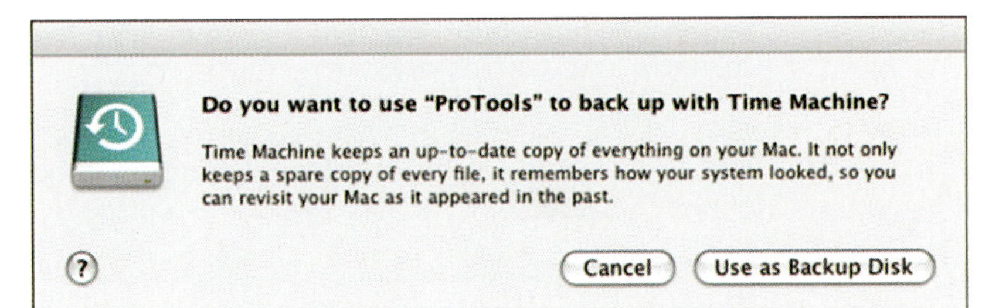

Figure 1.22 Turning off Time Machine with Pro Tools 8 LE.

For Windows systems, make sure you format your drives using New Technology File System (NTFS). You can use the Disk Configuration tool found in the Computer Management console. Chapter 7 covers backup and recovery of session data.

1.18 Installing and upgrading to Pro Tools 8 LE

Whether performing a fresh installation or an upgrade, there are a few general rules for upgrading any system. Before you upgrade, it's wise to organize your files. Remove unnecessary projects and get your recorded work situated and accounted for. This will make the upgrade process go more smoothly. You should always make a backup of your data. If you have an issue, you will not have to worry about losing your work. No matter what application you upgrade, it's always a good idea to back up any critical data on your system. Before performing the upgrade, always make sure that your system is compatible. It should meet the minimum requirements listed on Digidesign's Web site. To check your system's minimum requirements, go to the hardware compatibility matrix online on Digidesign's site (http://www.digidesign.com/compato/). Once there, search for the operating system you are currently running. Here, we will be focusing on the Power Mac G5 running the OS X operating system. To verify if you meet the minimum requirements, first check your system resources by clicking the Apple menu and then About this Mac. For detailed information, click the button labeled More Info. This is where you can find details about your system, and then check Digidesign's Web site to ensure that you are able to install and run Pro Tools 8 LE without a problem. You should also check to make sure that all your plug-ins (or other third-party software) are compatible with Pro Tools 8 LE and download updates if necessary.

When installing and running Pro Tools, make sure you are logged into your system using an account with administrative privileges. Failure to do so will result in error messages and problems starting Pro Tools 8 LE. If you find yourself not able to start Pro Tools, you may want to check the account you have logged in with because it may not have the correct privileges associated with it. When installing Pro Tools 8 LE, make sure you pay close attention to where you install it on your computer. Make sure that you install it to a drive with plenty of available disk space.

Next, if you have an executable, you can launch, whether on the desktop or on a disk, run the program to install, and upgrade to Pro Tools 8 LE. Figure 1.23 shows the installer's welcome screen.

The next few steps are pretty easy and ask you to agree with the Software License Agreement (SLA) and turn off the services we just discussed, such as Spotlight

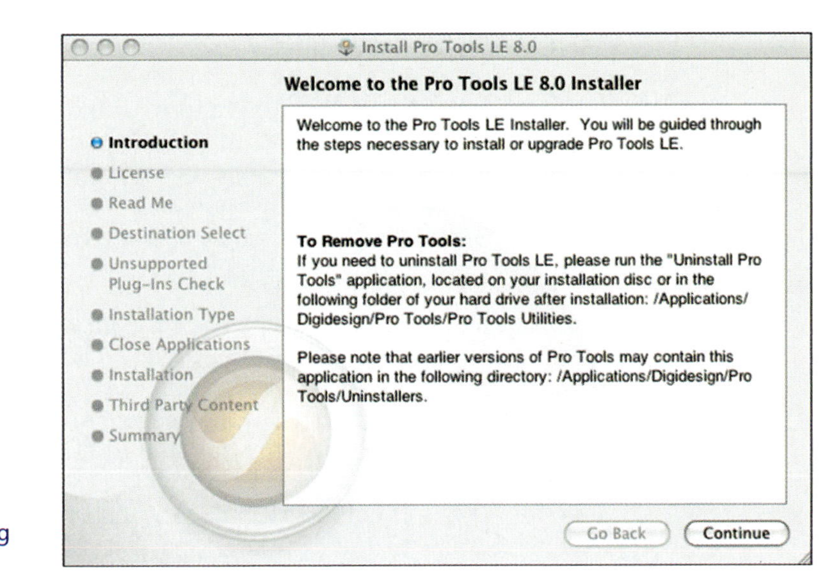

Figure 1.23 Installing Pro Tools 8 LE.

and Dashboard. You are then given the option to select specific components for your installation. Figure 1.24 shows the options available for installation.

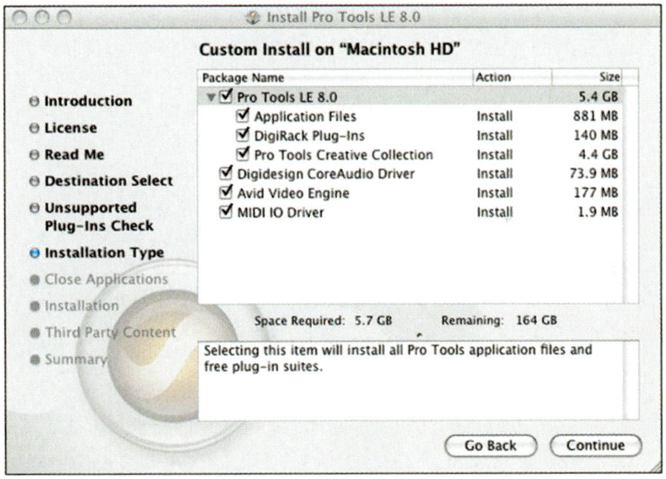

Figure 1.24 Customizing a Pro Tools 8 LE installation.

Be aware that tools such as the Avid Video Engine will need authorization codes, so make sure that you only select what you have purchased. If you select something you do not have access to, when you first open Pro Tools 8 LE for use, the system will hang up while trying to locate the missing information. Once you click on the Continue button, you will be told how much disk space your installation will take up and the option to start the installation will appear.

Your system will be checked to make sure that you have the space you need. You may even be asked to log in and allow the installer to continue. You are asked to close any running applications, and once you click Continue, the installation process begins. This usually does not take longer than about 10 min. If you experience any issues where it takes longer than that, you may not have enough disk space, which is why it's very important to ensure you do beforehand. Ending processes irregularly and terminating them incorrectly may corrupt your system, data, or software.

Once the installation process completes, you are asked by the installer to locate and install any third-party content found within your installation disk or download. Be aware that you can download these same installers from your MyDigi account at http://www.digidesign.com if you need to set up more content later. Make sure you reboot your system to complete the installation and then prepare to launch and use Pro Tools 8 LE for the first time. Launching Pro Tools and authorizing its use is easy. Once your computer has been rebooted, log in to the computer and then launch the Pro Tools 8 LE application. You can find and launch Pro Tools by going to the Go menu and then selecting Applications from the drop-down menu. Once you have Applications in view, you can find Pro Tools in the Digidesign folder. Locate the Pro Tools program icon and drag it to the Dock to make a shortcut for easy access. You can use the Dock to have quick access to Pro Tools. Once you launch Pro Tools for the first time, you must activate it correctly. You will need to enter the authorization code provided with your copy of the software. If this is a new copy, then you will find the code on the inside of your Getting Started guide. If you download your copy, you will receive copies of this information in your email. Figure 1.25 shows the authorization screen for Pro Tools 8 LE.

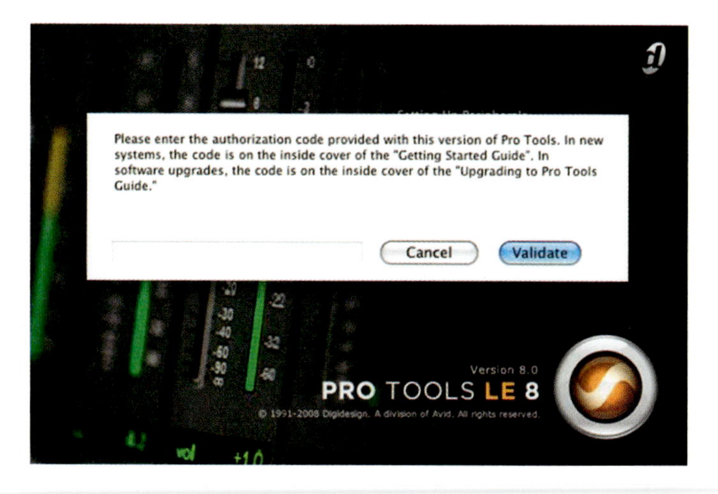

Figure 1.25 Configuring authorization for Pro Tools 8 LE.

Once you have validated your copy, it's ready for use. Now you can launch Pro Tools 8 LE and begin using it. If you are asked to register, as seen in Fig. 1.26, you can choose to do so at this time. If you already have (as seen in the graphic), you can bypass this request.

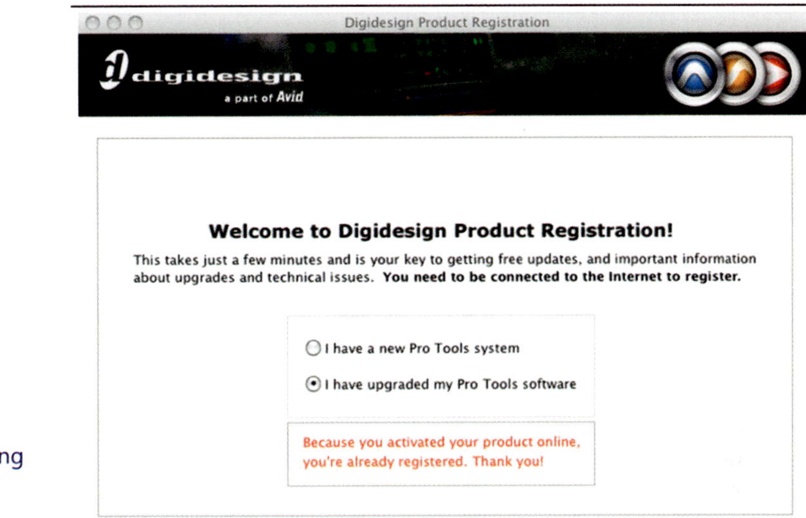

Figure 1.26 Registering your copy of Pro Tools 8 LE.

Lastly, if you had any issues getting your software, it's recommended that you set up a MyDigi account online to manage your installations and get access to your software online. You need a MyDigi account if you did not receive your authorization codes. Figure 1.27 shows a list of software you can download if

Registered Downloads

Downloads		
Pro Tools LE 8 Installer (Mac)	4152.55 MB	Download
Pro Tools LE 8 Installer (Windows)	3327.81 MB	Download
Pro Tools LE 8.0 Mac Read Me (PDF)	0.11 MB	Download
Pro Tools LE 8.0 Windows Read Me (PDF)	0.13 MB	Download
Upgrading to Pro Tools LE 8.0 — Installation Instructions (PDF)	0.45 MB	Download
What's New in Pro Tools LE & M-Powered 8.0 (PDF)	7.91 MB	Download
Big Fish Dance DJ Loops	1229.6 MB	Download
Big Fish Hip Hop Loops	1267.08 MB	Download
Big Fish Jazz Loops	516.01 MB	Download
Big Fish Pop Rock Loops	1094.18 MB	Download
Big Fish R&B Funk Loops	1226.28 MB	Download
Big Fish Rock Hard Loops	371.43 MB	Download
Big Fish World Loops	399.43 MB	Download

Figure 1.27 Selecting software or download on MyDigi.

you are authorized to do so. This gives you immediate access to software, and if you have your iLok key, all the licenses are kept safe.

Removing Pro Tools 8 LE

In some cases, you may need to remove Pro Tools from your system, such as if Pro Tools gets corrupted or does not run correctly. To remove Pro Tools 8 LE, run the installer from the Web download or CD-ROM. Select Uninstall and follow the prompts. Reboot your system once completed. Mac OS X users should Repair Disk Permissions before reinstalling Pro Tools. If you need to remove Pro Tools software from your Windows-based system, use the Add or Remove Programs command applet found in the Control Panel.

As you can see, setup is lengthy but very important. If possible, test your equipment prior to a recording session to make sure that everything is in working order and connected correctly. An incorrect setting or connection may not show up as a problem until you start recording. Troubleshooting while the artist is trying to record is not ideal. The process of prepping and checking your area will ensure that the recording process goes smoothly. When you prep the room, you are either checking the equipment already set up or setting up new equipment. You may have to configure the room differently based on the needs of the artist, which could introduce many new factors.

1.19 Powering the DAW on and off

To ensure that you do not damage any speaker components, always remember that you will want to turn on your speakers (monitors) last when powering up your system and first when powering down. You may also want to verify that volume settings are very low or down completely when beginning your session. When starting your system, you should also make sure that all components are in sync with regard to the sample rate and clock source.

To make sure that you power your system on correctly, follow these simple steps:

1. Turn on all external hard disks. Give the disks a few seconds to start up.
2. Power on all MIDI and synchronization devices you have connected to your DAW. Turn on your 003 or Mbox2 unit. Power on your control surface.
3. Power on your computer. Once booted up, log in and run Pro Tools 8 LE.

To power down your system, you would follow these same steps in reverse. Lower all volumes. Save your work, quit the Pro Tools application, and power down your computer. Turn off your external devices (such as microphone preamps) and then, lastly, your external hard drives.

Following these simple steps will allow for a smoother startup and shut down of your DAW. It will also extend the life of your components, reduce problems when launching and using Pro Tools, and lastly, save your speakers from turn-on thumps and the damage they can create. You should always take note of the volume level in your headphones and always take them off when powering down your system, as they can cause serious damage to your ears if the volume is not controlled.

1.20 Summary

In this chapter, we learned about a DAW, its components, how to select what you need to build your DAW, and how to scale it with extra gear and cable it. We then set up a Leopard-based system and installed Pro Tools 8 LE for use.

In the next chapter, we will cover Session Setup. Here, we open Pro Tools 8 LE for the first time and get to work on creating a session, using the new template options, making templates, basic navigation, and much more.

In this chapter

2

Session Setup

Now that your DAW is complete, you are ready to start using Pro Tools 8 LE and taking advantage of its many features. In this chapter, we start up a new session, learn about session parameters, using common tools, and navigating the Workspace. We also learn how to use session templates, using the Mix and Edit windows, transport tools, tracks, plug-ins, routing, input and output (I/O), and preferences. This chapter is a complete walk-through and gives you a great understanding of how to work with many of Pro Tools' most critical features. It will get you ready to begin the next phase of music production – composing your work!

2.1 Understanding Sessions

Now that Pro Tools is ready to go, you are ready to open your first session. Chapter 1's final section consisted of installing Pro Tools 8 LE and getting it ready for its first launch. On launching Pro Tools from its icon in the Dock (see Chapter 1 to locate the icon), you are greeted by a new Wizard (called Quick Start) prompting you to open a preexisting session or select other options such as making a new session from a preconfigured template. Before getting deep into what new options are available though, we should quickly cover the basics of sessions. Also, we need to make a few more adjustments to Pro Tools to prepare it for further use. Before composing, recording, editing, mixing, and final mastering take place, you need to know how to create a session, prepare it for use, and additionally know how to navigate and operate Pro Tools correctly. But before we get into the actual mechanics of opening a session and setting its parameters correctly, there are a few things that you should know about file management.

Sessions are nothing more than a set of files on your computer system where all session data (such as audio, video, and other) are stored and located for use. When you create a new session, Pro Tools will automatically create a new folder for your session that you can name to identify it. Within this folder are the session file, a WaveCache.wfm file, and several subfolders (an Audio Files folder, a Fade Files folder, a Region Group folder, and a Session File Backups folder). The session file is the document that Pro Tools creates when you start a new project. The session file has a *.ptf (Pro Tools file) extension. The Audio Files folder contains all audio recorded or converted during the session. When you record a new audio track, the track is saved as a new audio file to the Audio Files folder. You can also import other audio files into the session and work with them as well. Chapter 7 covers importing and exporting data in and out of Pro Tools in great depth.

The Fade Files folder contains any cross-faded audio data generated by the session. The Region Groups folder is the default directory for any region groups that you export from your Pro Tools session. The WaveCache.wfm file stores all the waveform display data for the session. If you delete the WaveCache. wfm file, Pro Tools will create a new one the next time you open the session. Storing waveform data in the WaveCache.wfm file allows the session to open quicker. The Session File Backups folder contains automatically generated back-ups of your Pro Tools sessions. These files are created when working on a session and the Enable Session File Auto Backup is enabled within the Operations tab of the Pro Tools LE Preferences dialog.

By creating a session, you are essentially able to group a project's data together and using the Workspace (Fig. 2.1), you can work with, link up, and even repair session file data. To get to the Workspace, go to the Window menu with a session loaded and select Workspace.

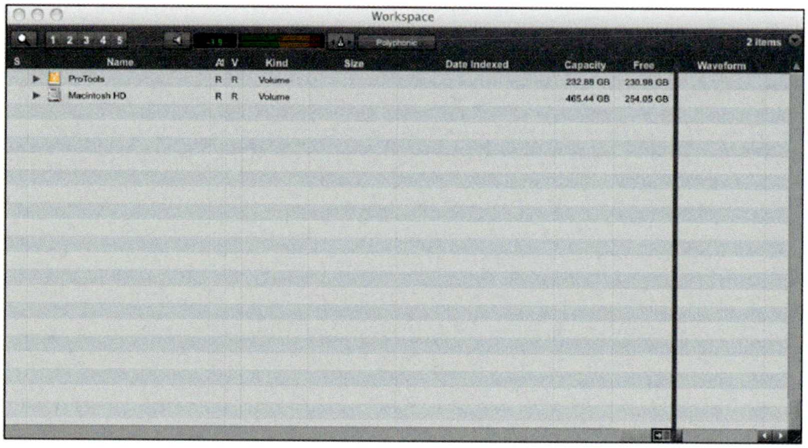

Figure 2.1 Viewing session data in Workspace.

You can do a lot within the Workspace, so make sure that you understand how to get to it and navigate it. The Workspace will be covered again later in this chapter when you finish configuring your drive for use with Pro Tools 8 LE.

Setting your drive up for recording

If your external drive is not set up correctly, you will receive error messages and likely not be able to record from it. The most common errors are (1) your selected drive cannot be designated as an Audio Playback volume because it is not a valid audio volume or (2) the selected drive is simply not an Audio Playback volume. Both error messages lead to the inability to record to the volume, as seen in Figure 2.2.

To fix this issue, simply adjust the settings in Workspace as described next. Before recording, any drive connected to your system must be prepared for use with Pro Tools. In the Workspace browser, make sure that the drive is designated for recording. To do so, click on the Window drop-down menu and click on Workspace. Drives may be set up to Playback, Transfer, or Record. To make sure that you are able to record data to your drives set them to Record or "R." If you wish to write protect a certain drive, the setting can be changed to Playback or "P." If you still get an error, you must reformat the drive correctly using the Disk Utility tool, covered in Chapter 1.

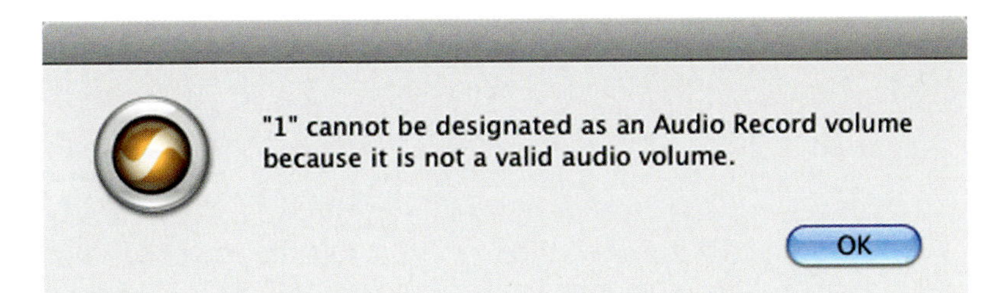

Figure 2.2 Audio Playback error.

2.2 Creating a New Session

The first step in beginning a Pro Tools project is creating a New Session. Launch Pro Tools and you will be greeted by the Quick Start dialog, as seen in Figure 2.3.

This wizard is helpful in guiding you on your way when using Pro Tools. One of the common complaints of users of older versions was that new users encountered a steep learning curve when getting started with Pro Tools or just about any audio editing software tool. Sure, audio engineering is titled as such because you do in fact need to study the subject to apply it correctly, but getting started and setting up a session with tracks shouldn't be as complex or time consuming as it had been. Now with Quick Start, you are literally on your way and running an entire session in minutes.

Figure 2.3 The new Quick Start dialog.

Pro Tools Quick Start dialog is used for quickly and easily creating new sessions from Digidesign's templates. There are multiple ways to create a new session and, as you can see from the dialog, you can create a session from a template, create a blank session (or new session without any preconfigured settings), open a recent session you had been working on (up to 10 entries in the history buffer), or simply open a session from a specific location on your hard disk. No matter which option you choose, you are not going to be able to do any composing, recording, editing, or mixing without a session setup. Be aware that if you do not see the Quick Start dialog, you can choose to hide or display the Quick Start dialog at launch in the Pro Tools Preferences menu, in the Operation tab.

If you create a new session from a template, you'll find that you have quite a few to choose from. For example, you will find a handful of guitar templates you can get started with immediately. Each template when opened will have approximately 10 or more tracks already in place to help you get started. Some of these tracks Include a Click track used to help set and keep your time and tempo, an Instrument track configured with a MIDI drum beat, tracks already configured with amplifier style plug-ins, and much more. These templates really help you not only get started but also understand basic track layout and configuration. You now have several working examples to choose from. Before Pro Tools 8, you needed to know how to configure a Click track for use – now one comes configured for you. This makes getting started much easier and makes the learning of Pro Tools much more fun.

There will be times when you want to have access to your own templates and you can now configure your own template set for your recordings. To do so, you need to know how to create a template file. You can make your own session templates quickly and easily and access them in the Quick Start dialog. As we mentioned before, sessions are nothing more than files. That being said, almost all files on computer systems have what is called a file extension. This extension is what tells the operating system what program or set of programs to associate with the file. For example, Pro Tools sessions use the extension *.ptf. If you saved a session named guitar_test, the session file would be saved as guitar_test.ptf and this file would be located in the folder in which you saved the session. Pro Tools Session Template files use the *.ptt extension. To create a new session from a template, simply select the option to Create Session from Template. Then, select which template you would like to work with. Click on the Session Parameters reveal button to adjust any session parameters that need to be adjusted such as Bit Depth and Sample Rate. Click OK and make sure you save the session once completed. Make sure you save the file with the *.ptt extension to create a template file.

To create a new session, you can use Quick Start or if Pro Tools 8 LE is already open, click on File, then New Session. You will open up a New Session dialog, as seen in Figure 2.4.

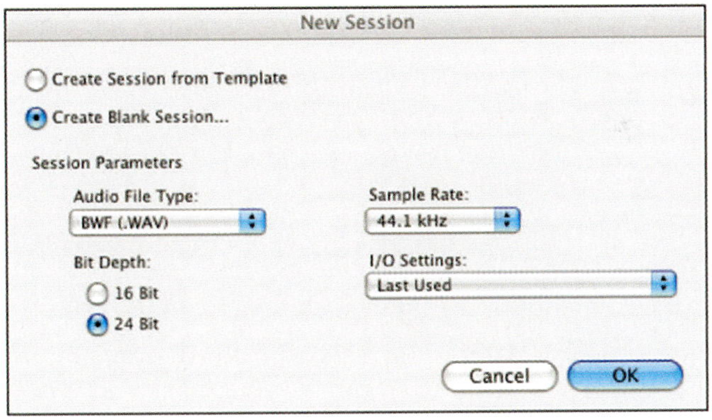

Figure 2.4 Creating a new session.

If you want to change any of the session parameters, click on the Session Parameters reveal button. Here, you can set several different options. Next, select the audio file format for the session. To ensure compatibility between Windows and Macintosh platforms, set the file type to BWF (.WAV). Windows systems do not support Sound Designer II (SD II) files. For this reason, WAV files are usually preferred when moving sessions between different Pro Tools systems. SD II files also do not support sample rates above 48 kHz.

In this window, you are also able to select the Bit Depth (16 or 24 bit) and the Sample Rate (44.1, 48, or 96 kHz). CD resolution is 16 bit, 44.1 kHz. For a higher fidelity and more detailed sound quality, many engineers prefer to use a higher bit depth (24 bit) and sample rate (48 or 96 kHz) for recording, and then convert the signal down to CD resolution. When selecting a Bit Depth or Sample Rate for your session, consider fidelity, any compatibility issues with others systems, and storage space. Higher resolution settings take up more of your computers resources. As bit depth and sample rate increase, the number of tracks and plug-ins you can use without dropouts decreases. In addition, files of different bit depths and sample rates than the current session must be converted to the current session's settings before they can be imported into the session.

Next, select the I/O settings to use for the session. Choose the Last used if you are opening Pro Tools 8 LE for the first time. Finally, name the session and click on Save. When opening an existing session, simply choose File, click on Open Session, locate the session you wish to open and click Open. Saving a session is equally simple: choose File, Save Session, then give the session a name and click Save. From now on, your session will be overwritten each time you save. If you wish to save a session with a different name, use the Save As command in the File drop-down menu. If you wish to save a copy of the session, click the Save a copy in command from the File drop-down menu, name the session, and click Save. Save commands cannot be undone. As noted before, Pro Tools sessions should be saved on a dedicated audio drive if possible. Usually it's best to save the session data (edits and templates) on the system drive, and save the audio files on a separate drive so that they can stream while playing without interruption.

2.3 Configuring Hardware Setup

Depending on the components connected to your system, you may have to make changes to the Hardware Setup of Pro Tools 8 LE. Clock source and synchronization are very important to the stability of your system as well as the overall sound and quality of your recording. Depending on your system configuration, the clock source may need to be set. If you have external devices connected through the ADAT Optical or S/PDIF inputs, the clocking should be assigned to the external device. To find the clock source, go to Setup menu, click on Hardware, and set the clock source to ADAT or S/PDIF to match the type of connection you are using. When connecting to devices via the ADAT optical (003 only) or S/PDIF digital connections, setting the clock source to Internal may often cause pops, clicks, and distortion, but if there are no external devices connected via these connections, the Internal setting should be used. When connected to another time source via the ADAT connection (over fiber), the 003 will seek a Master Clock source from the other device. You need to configure the other source as the Master Clock or it will not work. In addition, other hardware-based settings such as the digital I/O format and the default sample rate can also be changed in the Hardware Setup window. To check your DAW for correct configuration settings when using external ADAT devices, refer to Figure 2.5 (for an Mbox2 Pro) and Figure 2.6 (for the 003 Rack).

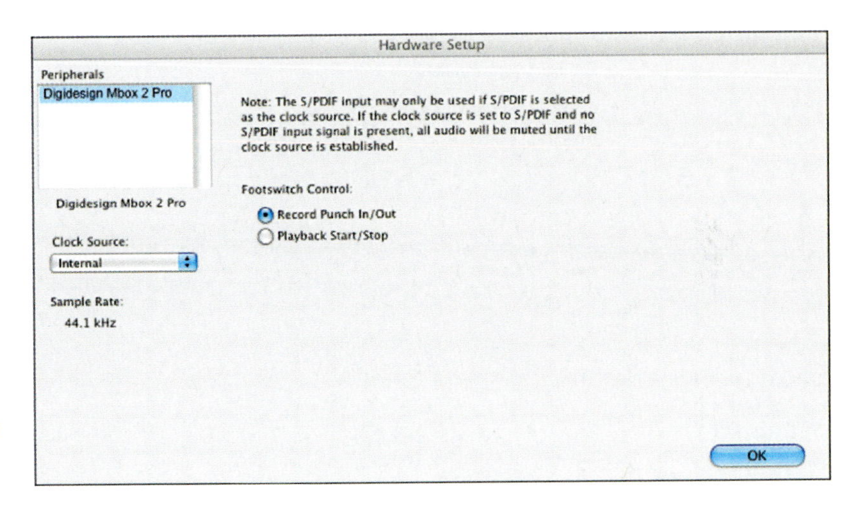

Figure 2.5 Configuring Hardware Setup for the Mbox2 Pro.

Figure 2.6 Configuring Hardware Setup for the 003 Rack.

2.4 Configuring the Playback Engine

To make the most efficient use of Pro Tools, you may need to adjust the way your computer performs while performing specific operations. From the Playback Engine window, you can adjust the performance of your system by changing settings that affect its capacity for processing, playback, and recording. You may set the Hardware Buffer Size, the number of processors used, and the usage limits. The Playback Engine dialog can be found by clicking on Setup and then Playback Engine. The Playback Engine is seen in Figure 2.7.

Figure 2.7 The Playback Engine.

The default settings for your system may provide acceptable performance, but in most cases, you may want to adjust them to reduce latency (set a smaller buffer size) or to accommodate large or processing-intensive Pro Tools sessions (set a larger buffer size). The amount of Real-Time Audio-Suite (RTAS) plug-ins and other audio processing Pro Tools can handle per session is dependent on the amount of CPU power and other hardware resources available to Pro Tools.

The Playback Engine dialog is helpful to any engineer looking to work with Pro Tools LE efficiently, as by understanding this dialog and knowing how to operate it; you will get more power out of your Pro Tools 8 LE system. Figure 2.8 shows a commonly found Playback Engine error.

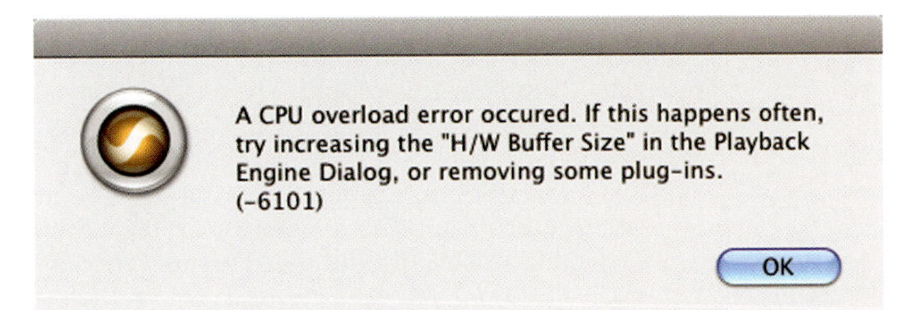

Figure 2.8 A common overload error.

For example, if you are recording a soft synth, you should use a lower H/W Buffer Size to reduce latency in the monitored signal, whereas you may want to set a high H/W Buffer Size (let's say 1024 Samples) for mixing and playback to give more power to your digital plug-ins.

The CPU Usage Limit controls the maximum amount of CPU resources dedicated to Pro Tools. The maximum CPU Usage Limit is 85% for single-processor computers (except for 003, which has a limit of 99%) and 99% for multiprocessor computers. On multiprocessor computers, the maximum CPU Usage Limit is reduced when you use all the system processors. For dual-processors, the limit will be 90%. On four-processor computers, the limit will be 95%. The 99% setting dedicates one entire processor to Pro Tools. In most cases, leaving a small percentage of CPU power for other system tasks is desirable. When running other applications at the same time as Pro Tools, leaving some headroom may help guard against a system crash. Large sessions or using more real-time plug-ins may require a higher setting, but note that increasing the CPU Usage Limit may slow down screen responses on slower computers.

Latency is the delay that occurs when your software processes audio and sends it to your speakers. It adds a delay to the monitored audio. Latency is directly affected by the Hardware Buffer Size. The Hardware Buffer Size (H/W Buffer Size) controls the size of the hardware cache used to handle host-based tasks such as RTAS plug-in processing. Lower H/W Buffer Size settings reduce monitoring latency and are useful when you are recording a live source. Higher H/W Buffer Size settings allow for more audio processing and effects but may cause slower screen response and increased monitoring latency. Higher settings are useful when you are mixing and using more RTAS plug-ins. In addition, Hardware Buffer Size settings can affect the accuracy of plug-in automation, mute data, and timing for MIDI tracks. If you are experiencing slow system response, artifacts in the audio, or error messages, the H/W Buffer Size may need to be increased.

The RTAS Processors setting determines the number of processors in your computer allocated for RTAS (real-time plug-ins) plug-in processing. With computers that have multiple processors or that feature multicore processors or Hyper-Threading, this setting lets you enable multiprocessor support for RTAS processes. Choose one processor to limit RTAS processing to one CPU in the system, or choose two processors to enable load balancing across two available processors. On systems running four or more processors, choose the desired number of RTAS processors as needed. When using Pro Tools, it is usually preferable to use the maximum number of CPU processors in your system.

You can also tweak your memory usage although this is not always recommended. For example, you can adjust the DAE Playback Buffer to different level settings to set the amount of memory Pro Tools will set aside for its playback buffer. You can adjust the Size (Level 0, 1, 2, or 4) and the Cache Size (minimum, Normal, and Large). Although this setting is not adjusted all that much in production environments, it helps to understand when and why you would adjust it, as knowing how to adjust and work with system memory and other system resources can help you be more productive. As mentioned earlier, your hard disk is where session data is stored and used, so it's important to have a faster path to and from the disk. Pro Tools takes audio and other session data from the hard drive and then into the DAE Playback Buffer. If you are setting the Size higher than Level 2 to get better response time and less latency, you may need to do the following: backup your session data and reformat your drive, upgrade your drive (external FireWire 400 or 800) type, or speed (7200 RPM drives or higher recommended), which may solve your latency issues. A seriously fragmented drive or a drive that is too slow for Pro Tools to use should be repaired and/or replaced if the issues cannot be resolved. As noted before, because a system disk can be a bottleneck, it is recommended that you use an external FireWire disk or a separate 7200 RPM internal disk for audio files to achieve the least amount of latency and best performance.

It's important to understand how to use the Playback Engine because when running Pro Tools 8 LE, you may find that you receive an occasional error or two asking you to adjust the Playback Engine for better results. An H/W buffer error is a common occurrence on DAWs pushing dozens of tracks using plug-ins, routing your signal in and out and so on. Understanding and working with the Playback Engine will help you get the maximum capability out of Pro Tools 8 LE when power consumption is at its highest.

2.5 Configuring the I/O Setup

I/O (Input/Output) is the terminology used to describe the path which a signal is routed through Pro Tools. It's extremely important to understand I/O Setup and operation because in later chapters when you start to record, edit, mix, and master, the routing of your signal becomes crucial to your projects. The I/O configuration lays out the "roadmap" for the signal to follow. When using Pro Tools for the first time, Pro Tools automatically sets a default configuration for I/O Setup. If you are using external hardware, these settings may need to be changed according to your needs. The I/O Setup dialog also lets you save and import I/O setting files. Your configuration can then be quickly applied when setting up new projects. The I/O Setup dialog as shown in Figure 2.9 provides a way to label, format, and map Pro Tools Input, Output, Insert, or Bus signal paths for each session. Since we are looking at an Mbox2 Pro, there are two inputs (left and right stereo or two monos) and the Stereo S/PDIF connection.

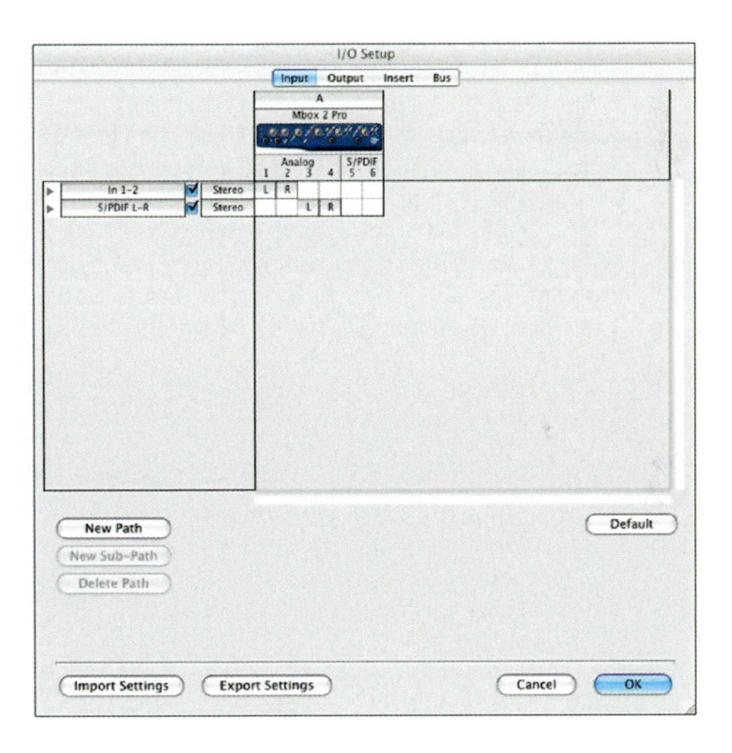

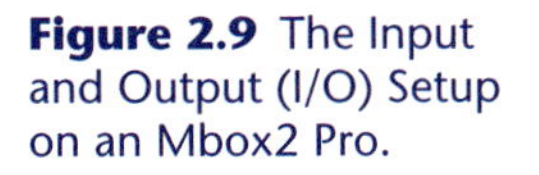

Figure 2.9 The Input and Output (I/O) Setup on an Mbox2 Pro.

When navigating through the I/O Setup dialog, you will be able to adjust Input, Output, Inserts, and the overall Bus. In Figure 2.10, we look at the Output section. This Output section is used to send audio to a master fader, headphones, or monitors.

Figure 2.10 The stereo output.

Figure 2.10 displays the default settings for the most common routing of your output signal within Pro Tools. If you wish to change the audio routing of Pro Tools, configure external hardware such as outboard gear for use, or create new sections and label headings, open the I/O Setup, and click the Input or Output tab to display the corresponding path type. To change a setting, click the Input or Output selector for the first interface channel pair located below the first audio interface icon. From the pop-up menu, select a physical port pair (such as Analog 1–2) to route to a Pro Tools channel pair (such as A 1–2) in the Path Name column on the left. Once this step is completed, repeat the above step for additional channel pairs. Click OK to accept and apply your new I/O settings. All paths must be valid before the I/O Setup configuration can be applied. To adjust or modify insert paths, simply click on the Insert tab in I/O Setup. Figure 2.11 shows common settings for the Insert tab.

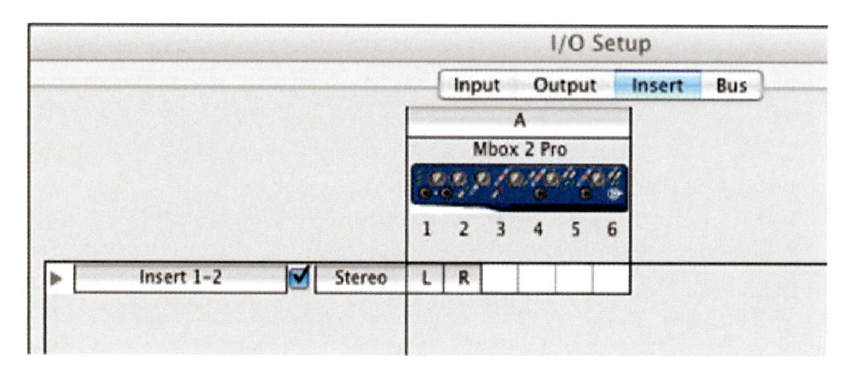

Figure 2.11 An Insert path in I/O Setup.

Selecting the Bus tab will help you configure the entire Bus layout, which can be accessed from multiple places within the Mix and Edit windows in Pro Tools 8 LE. Figure 2.12 displays the Bus section in I/O Setup.

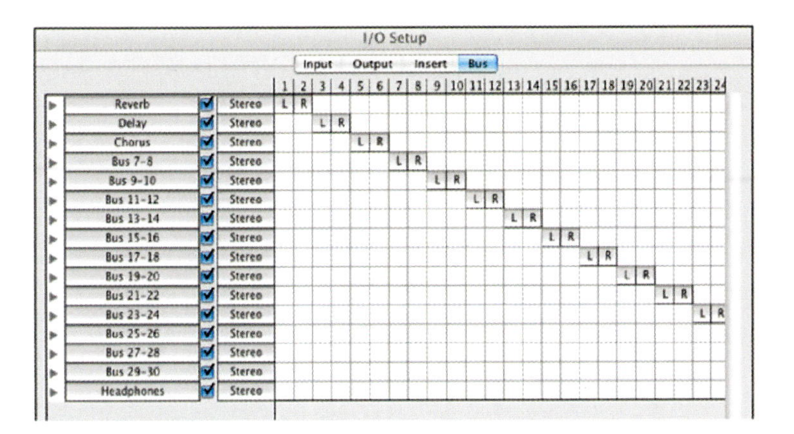

Figure 2.12 The Bus section in I/O Setup.

Under the Bus tab, you create the framework for where you will send your signal from multiple sections within your Pro Tools session. As seen in Figure 2.12 (as an example), you can change the I/O for a specific effect such as Reverb. When you create an auxiliary track later in this chapter, you can set up the routing of it with specific channels. This not only helps with organizing your routing infrastructure but also allows you to configure specific sections of your bus for specific tasks. You can also choose to set up a New Path, Sub-Path, or Delete Path. Each option will allow you to alter your paths while working within Pro Tools 8 LE. The Default button returns you all your original default settings, so you can experiment with the base configuration without worrying about losing your I/O Setup that works.

2.6 Configuring MIDI Devices

As mentioned in Chapter 1, you have the option to connect MIDI devices to your DAW. You need to make further configuration changes within Pro Tools if you add or change these devices. When connecting MIDI devices, use the Apple Audio MIDI Setup (or Apple AMS) window to verify that your MIDI devices are connected correctly and are routed properly between each interconnected device and then to Pro Tools. To do so, open the AMS window in OS X by clicking Setup, MIDI, and then MIDI Studio. In this window, correctly connected MIDI interfaces automatically show up with each of their ports numbered in a chain sequence. To set up MIDI devices connected to the MIDI interface, click Add Device to create a new icon for the device. You must then connect the new device to your interface by drawing a new connection or "Cable" using the arrow of the corresponding port of the MIDI interface. Connect both inputs and outputs and repeat for each additional MIDI device you may have. Click Show Info and choose the appropriate manufacturer and model of your device. From there, you may also specify the channel that the device operates on. Once connected, your MIDI device will be ready for use.

You should be aware that you may experience problems when using MIDI devices. If Apple's AMS application is launched and left running before launching Pro Tools 8 LE, you may experience a problem or receive an error message. If this is the case, test and configure your MIDI devices again and then close the AMS before launching Pro Tools 8 LE.

2.7 Configuring Pro Tools Peripherals

If you wish to configure additional peripherals select the Setup menu in Pro Tools and then choose Peripherals. Once selected, you will be given the option to choose between further menu options such as Synchronization, Machine Control, MIDI Controllers, and Ethernet Controllers.

The first tab, Synchronization, allows you to adjust the MTC Reader and Generator ports. Apogee Big Ben Master Clock or Rosetta products that do Synchronization via MIDI may require that you adjust specific synchronization settings for the systems to work together. Figure 2.13 shows the Synchronization options.

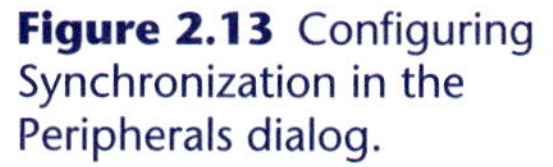

Figure 2.13 Configuring Synchronization in the Peripherals dialog.

The next tab is Machine Control, which is used to denote a Master MIDI Machine Controller. This is where you can set the master and configure its MIDI ID number. You can also enable and configure a Slave Controller for use. You can use these settings to configure a slave out of another audio editing suite such as Logic Pro or Cakewalk Sonar.

Next, as seen in Figure 2.14, you can configure a MIDI Controller for use. This would be where the Command|8 Control Surface would appear when installed.

Figure 2.14 Configuring a MIDI Controller.

In some cases, you may need to adjust the Receive From, Send To, and # Ch's settings for proper routing. The Ethernet Controllers tab is where you can configure the C|24 Control Surface for use. In Figure 2.15, you can enable the Ethernet port (en0) and configure your surface as needed. You can also adjust the color coding as well.

Figure 2.15 Configuring a new control surface (the C|24).

Once you have your peripherals configured, the next step is to start creating tracks and learning how to use them. In the next section, we will learn how to make tracks and work with them in the Mix and Edit windows of Pro Tools 8 LE.

Stereo versus Mono

Simply put, stereo means that although you have created one track, you are reproducing sound for two separate output devices – a left and right speaker pair (as an example). Using stereo means that you place (pan) each recorded instrument in a specific location between your two speakers, simulating how real instruments are spread out in space. Stereo recordings require the use of two speakers, two microphones, and left and right channels of audio. Time and sound pressure level differences between the left and right channels allow you to perceive which direction the sound is coming from.

A stereo file is twice the size of an equivalent mono file because a stereo file contains twice the information. If you are recording audio for television, you may not need large amounts of disk storage space for session files. Mono is a common recording choice when creating audio for television broadcast. Mono refers to a single channel of audio and requires the use of a single microphone or speaker. Mono is usually the left channel of a stereo pair when stereo is not available. If you send a mono (single channel) signal to two speakers, the sound is still mono, not stereo, because both speakers are playing the same signal.

2.8 Understanding Tracks

Now that your system is properly configured, you are ready to use Pro Tools to begin the creative process – composing and recording audio. To record anything, you need to know how to create and operate Pro Tools 8 LE tracks. There are five basic types of tracks, which are Audio tracks, MIDI tracks, Auxiliary Input tracks, Instrument tracks, and Master Fader tracks. We look at each in detail. Table 2.1 shows the breakdown.

All of the track types serve specific purposes. For example, an Instrument track can be used to create a Click track. A Master Fader track can be set as a stereo master volume control with the proper routing. First, create a stereo Master Fader track and set the outputs of any other tracks you wish to route to the Master Fader to the main output path, which is usually Analog 1–2 by default. The Master Fader output should point to the stereo output of your audio interface. Master Faders can also be used to control and process output mixes, monitor and meter an output (such as a bus or hardware output) to guard

Table 2.1 Track types and functionality

Track type	Functionality
Mono or Stereo Audio Track	A mono track usually contains a single voice. You can control the volume and panning of a mono track. A stereo audio track is a track that plays two channels of audio as a stereo pair. Typical stereo tracks are drum overheads, a piano in stereo, or a synthesizer.
MIDI Track	MIDI tracks store MIDI note, instrument, and controller data. You cannot select a track format when you create a MIDI track because audio is not recorded on a MIDI track. You can select the type of timebase for the track (samples or in ticks).
Mono or Stereo Instrument Track	Instrument tracks are a special type of track that can provide MIDI as well as audio capabilities in a single track. Instrument tracks when combined this way can provide for simplified routing of anything from virtual instruments to MIDI sound modules. They are used to simplify the use of software and hardware instruments to record MIDI as well as to aid in the monitoring of audio when in playback.
Mono or Stereo Auxiliary Input Track	Auxiliary Input tracks (also seen as Auxiliary Inputs or Aux Input) are the equivalent of effects busses or group busses in a hardware mixer. They can be used as part of an effects loop, destinations for submixes, a bounce destination, inputs to monitor or process audio (such as audio from MIDI sources), or for many other audio-routing tasks. You have the option to set up the Auxiliary Input track in multiple ways. The most common way to set up such a track is to create an effects bus. For example, insert a reverb plug-in in an Auxiliary Input track, then create a send to that auxiliary reverb track from a vocal track.
Mono or Stereo Master Fader	Master Fader tracks (or Master Faders) control the overall level of the audio tracks that are routed to the session's main output paths. For example, you could have 24 tracks in a session with channels 1–8 routed to Analog Output 1–2, channels 9–16 to Analog Output 3–4, and channels 17–24 to Analog Output 5–6. You could then create three master faders, one to control each of these output pairs. Master Fader tracks have additional uses (such as controlling submix levels).

against clipping, control submix levels, control effects sends levels, control submaster (bussed tracks) levels, and apply dither or other inserts to an entire mix.

When first learning to view tracks, it is easy to get confused because you can view "tracks" in both the Mix and Edit windows. Actually, in the Edit window, you are viewing tracks. In the Mix window, you are viewing channel strips. They are not the same thing. Channel strips are used to control the sounds that are recorded in tracks. A channel strip ("channel") is often used interchangeably with "track" and "voice." If you are going to record a singer and an acoustic guitar, at bare minimum you would need two tracks and two corresponding channel strips. In the Mix window, you would assign two channel strips, one for the vocals and one for the acoustic guitar. Later in the chapter, we navigate the Mix and Edit windows, so you can view how the tracks and their channel strips are laid out within them.

Samples and Ticks

Each track created in Pro Tools 8 LE has a "timebase" associated with it. A timebase defines where the audio and/or MIDI data is placed on the session's timeline. In the next few chapters, we will become more familiar with the timeline and how to run a session, but for now, it is enough to know about the two different timebase settings in Pro Tools 8 LE, sample and tick.

A sample is defined as absolute, whereas a tick is defined as relative. Ticks are commonly used with MIDI recordings, and samples are usually used in audio recordings. Samples that function on the absolute timebase are built upon the current sample rate, such as 44.1 kHz as an example.

2.9 Creating Tracks

To create a track with Pro Tools 8 LE, go to the Track menu and select New. You can now select any number of tracks to create with the New Tracks dialog seen in Figure 2.16, which contains a cross-section of the most common tracks you will create while working with Pro Tools 8 LE.

New Tracks

Create	1	new	Mono	Audio Track	in	Samples
Create	1	new	Stereo	Audio Track	in	Samples
Create	1	new	Mono	Aux Input	in	Samples
Create	1	new		MIDI Track	in	Ticks
Create	1	new	Mono	Instrument Track	in	Ticks
Create	1	new	Mono	Master Fader	in	Samples

Cancel Create

Figure 2.16 Creating multiple tracks.

Click on the Create button to get started after entering the appropriate track information. Once you do, Pro Tools 8 LE will create the tracks you requested and open them within the Mix and Edit windows as seen in the following figures. While you get started with your tracks, take a good look at the difference between the Edit and Mix windows. In Figure 2.17, you can see your new tracks opened up within the Edit window.

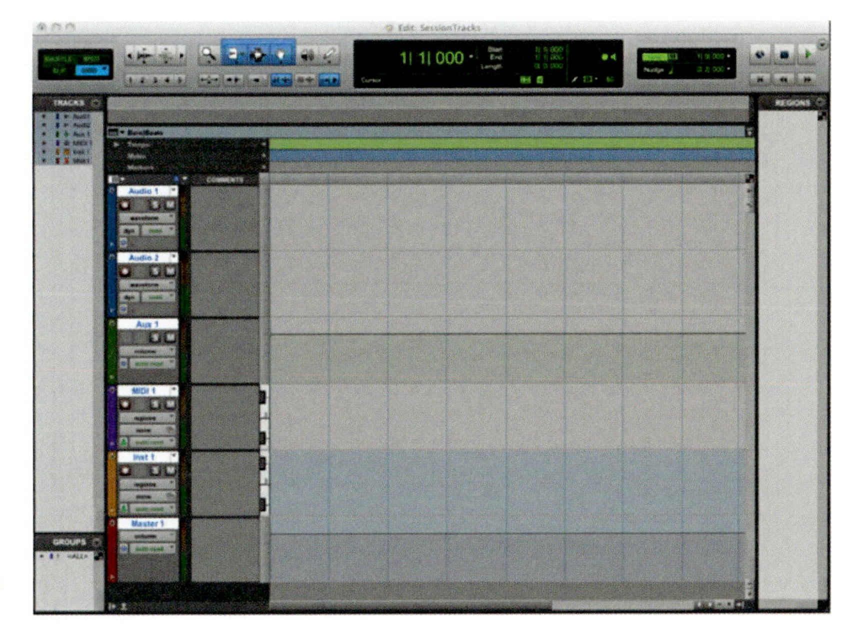

Figure 2.17 Viewing open tracks within the Edit window.

The Edit window displays waveforms that might be separated into Regions (segments of audio) within each track. Once you start to record or import audio and MIDI data, this window provides you with the tools and transport options to edit that data. Although we cover the Edit window in greater detail in Chapter 5, you need to know its basic functionality now to get your tracks loaded and configured.

The Mix window, as seen in Figure 2.18, is the other main window within Pro Tools and is where you will do most of your routing, mixing, and mastering. Although you can also use Mix functions within the Edit window, most audio engineers and producers work with both windows simultaneously to maximize their efficiency.

In the Mix window, you can see the channel strips for the tracks we just created and viewed within the Edit window. A closer look at each channel strip shows that each has many settings to adjust and configure. Editing will be covered in more detail in Chapter 5 and mixing in Chapter 6, but we will cover some of the more basic configurable options here. You

Figure 2.18 Viewing open tracks within the Mix window.

can refer to Table 2.2 for some of the track options you will be able to work with.

We will cover the specific portions or section of each track you use for recording, composing, mixing, and mastering later in this book. For now, you should feel comfortable with opening Pro Tools 8 LE, making a session, creating tracks, and getting to work. If you opted to use a preconfigured session template, you will find most if not all these tracks located within it. This is a great way to help you learn how to work with tracks in Pro Tools 8 LE. In this section, we briefly covered many of the different things you will do with tracks and many of the things you can adjust and work with. As we get deeper into the book, we learn in more detail about how to use and manipulate tracks.

Table 2.2 Common configurable track options	
Inserts	This is where you can add plug-ins, usually for recording, mixing, and mastering. By clicking on the up/down arrow you can select whether to add or remove Inserts such as Digital Signal Processors (DSPs). Pro Tools 8 LE allows for up to 10 inserts per track.
Sends	Sends are used to "send" the signal to an additional channel or bus. For example, you may be working with a track and need to send the signal to an Auxiliary Input track (effects bus). For example, suppose you have a reverb effect setup in an Auxiliary Input track. In each track that you want reverb on, create an effect send to the Auxiliary Input track that contains the reverb plug-in.
Audio Input/Output (I/O) Path selectors	You can use the Audio I/O Path selector to choose where to route your signal on the bus, similar to the Sends option. These selectors allow you to assign the appropriate interfaces to both receive and transmit audio through the proper routes.
Solo/Mute buttons	The Solo button will allow you to play only that track you have selected. You can set multiple tracks to solo at one time. Select Mute to mute any track while it is playing – this will render it inaudible.
Automation Mode selector	The Automation Mode selector is used to turn on Automation features. Chapter 6 covers mixing and automation in detail. By selecting automation, you can have all your session's track adjustments vary continuously and you can program what you want to happen through the session. This selector allows you to turn on (or off) different modes such as On, Off, Read, and Write.
Pan slider	Pan sliders allow you to pan a track's sound anywhere between your two monitor speakers. You will learn panning techniques when we cover mixing. For example, you can send your piano to the right side of your speaker pair and the guitar to the left. You can pan the drums left and right and send the vocal straight down the center of the mix.
Pan indicator	The Pan indicator shows you (with numbers) how far to the left and right your panning location is.
Track Record Enable button	When you select the Track Record Enable button, it allows the track you select to be recorded on. That means if you did not copy your currently saved track, you will record over it. When covering recording in Chapter 4, we will cover this in great detail.
Volume fader	This is used to control the volume level of the entire track.
Level meter	This meter helps you monitor the level of the track.
Voice selector	This gives the track (or voice) the ability to be active or inactive.
Group ID	This option lets you configure and set the Group ID. Grouping tracks is covered in Chapter 5, Editing.

(*Continued*)

Table 2.2 *(Continued)*

Track Type indicator	The Track Type indicator shows you what type of track you are working with.
Volume/Peak/Delay indicator	This indicator can be switched to show the track volume (input level) as well as peak and channel delay by pressing Ctrl- or Command-click to change between modes. The Volume/Peak/Delay indicator is crucial to recording and mixing your work.
Track Name	This section is used to note the track name. It is very important to label tracks correctly because eventually you could be working with sessions that could have up to four or five dozen different tracks!
Track Color Coding	Track Color Coding is extremely helpful in quickly identifying your tracks and track types.
Track Comments	Track Comments are also very helpful – these are used to store additional information about the track you are using. An example would be if you used an audio track to record a bass guitar. You may want to use the Track Comments section to note the type of bass guitar, amplifier, cabinet, plug-in, microphone, or other equipment used.
Track Position Number	The Track Position Number shows you the track number as it sits within either the Mix or Edit windows.

How to get more tracks out of Pro Tools 8 LE

When you need more track count, you can purchase it from Digidesign. The Complete Production Toolkit Software Option (Pro Tools LE Only) allows for expanded functionality such as increased track count and surround sound and panning features for mixing. If you do not need this additional functionality, then you can continue to read on without worry. If you are using Pro Tools 8 LE in a production project or professional studio, then you may need to consider getting this additional functionality. The expanded toolkit allows for an increased track count, with up to 128 mono/64 stereo at 48 kHz, 96 mono/48 stereo at 96 kHz, and 64 Instrument tracks. Pro Tools 8 LE has native support for up to 48 tracks (mono or stereo) giving you up to 96 available voices, whether configured as mono or stereo, at 48 and 96 kHz.

Additional features include the ability to move data between both Pro Tools HD and Pro Tools LE systems without any issues, as well an expanded software selection including the DV Toolkit 2 for video.

Caution

> You can delete a track easily by highlighting the track you want to delete, going to the Track menu, and selecting Delete. When working with Pro Tools, you will create and delete tracks frequently as you work. It's important to know that this action cannot be reversed. When deleting tracks, carefully note which tracks you have highlighted.

2.10 The Edit and Mix windows

Now that your tracks are in place, let's take a brief look at the two main windows you will use to record, edit, mix, and create audio. In Figure 2.19, you can see the Edit window with a session template loaded.

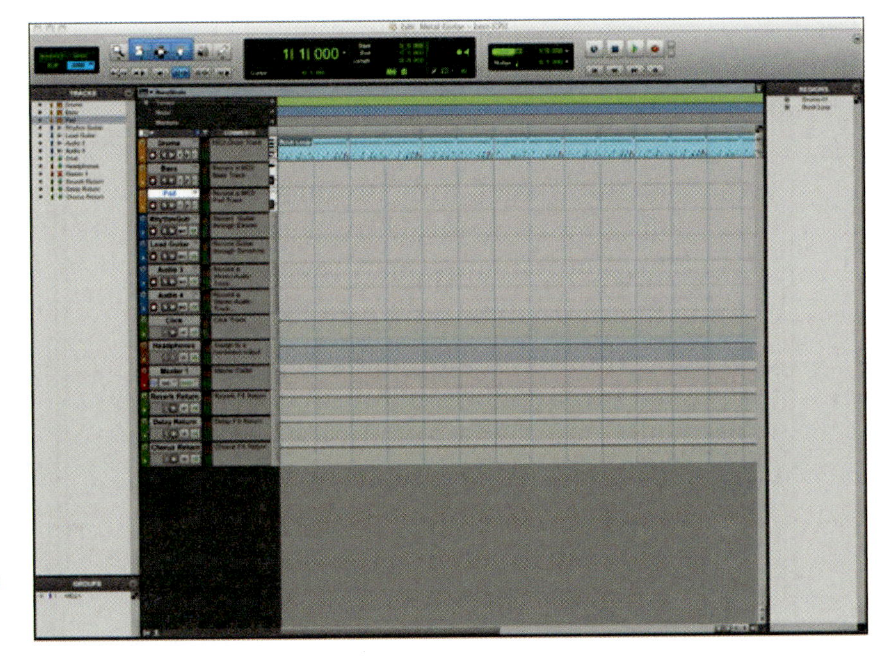

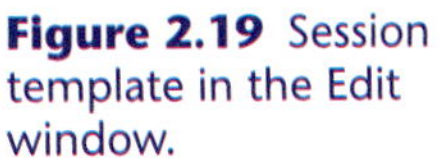

Figure 2.19 Session template in the Edit window.

When you first start a new session, you will be presented with the Edit window. If it later becomes minimized or is not in view, the way to open it is to go to Window menu and select Edit. As mentioned earlier, the Edit window is your primary window. Some of the biggest changes from the older version of the Edit window in Pro Tools LE 7.4.2 are a dramatically different color palette, the ability to configure more tools for use within the window itself and new track features that can be used when looping, recording, and so on.

There are many new features found in the Edit window, such as the new "Universe" view, new MIDI Editor view, new Waveform views, a customizable toolbar with the ability to add new sections to it, and the option to have Region Edit and Time Locking, as well as the Timeline Insertion follow the Playback indicator. Here, you can pretty much do anything and everything – in fact, if you customize your Edit window correctly, you can set up all your inserts, send, I/O, etc. right in the Edit window. While everything can be set up in the Edit window, many people, especially those with dual-monitor setups, prefer to put the Edit window in one monitor and the Mix window in the other to get the best of both worlds. You will see that once you start opening up other tools and programs, your virtual-desktop real estate diminishes quickly.

The Window menu is also where you can access the Mix window. Figure 2.20 shows the Mix window as well as the Edit window running on Leopard.

Note

Another window you want to familiarize yourself with is the Transport window that opens up when you first open your session template. This can also be loaded from the Window menu, by selecting Transport.

Figure 2.20 The Mix and Edit windows in Pro Tools 8 LE.

Before you get started composing and recording your MIDI and audio data, you should know how to work with the menus, preferences, and other key settings within Pro Tools 8 LE. This way when you move through the rest of the book's chapters, you can be assured that you know how to adjust the basic settings found within. Most of these configurations are found within the Pro Tools 8 LE interface and menu system, as well as the two main windows just mentioned – Edit and Mix.

When opening Pro Tools for the first time, you will find that there are many adjustments you want to make to the default installation. If you are new to Pro Tools, we will highlight some of the customizations you need to make.

2.11 Navigating Pro Tools menus

Pro Tools 8 LE comes with advanced window management tools; you can adjust how your windows sit on your screen (tile, cascade, etc), for example. To make all these adjustments, you should be familiar with the many settings you can work with and adjust within Pro Tools 8 LE. Table 2.3 shows the menu options in great detail.

Table 2.3 Viewing menu settings and options

Menu	Options
File menu	The File menu options include creating a new session or opening an existing session, closing or saving sessions, saving as a template, sending via Digidelivery, bouncing options, import and exporting options, and the ability to do scoring. If you have Sibelius installed, you can also send your session data to Sibelius.
Edit menu	The Edit menu options include undo and redo options, editing options such as cutting, copying, pasting, and clearing, selection options, fading tools, region consolidation, as well as silencing options (Strip Silence).
View menu	The View menu options include setting how the Mix and Edit windows can be displayed, and configuring rulers, guides, waveforms, sends, plug-ins and counters.
Track menu	The Track menu options include making new tracks, grouping tracks, track options such as deleting and duplicating, MIDI track options, scrolling options, and the setting for enabling an Instrument track to be a Click track. Another very important setting is Input Only Monitoring. During recording and playback, you want to be able to toggle between what you hear and what you playback when recording on a track. If Auto Input is selected, you will be able to listen to what is on the track until you start to record on that track. Then you will hear what is being recorded.

(Continued)

Table 2.3 *(Continued)*

Menu	Options
Region menu	The Region menu options include region grouping and editing functionality, looping functions, quantizing a selection to a grid, configuring elastic properties, and conforming to tempo. As you work through Chapter 3 and 4, we will use these tools and features in more depth.
Event menu	The Event menu options include configuring time and tempo settings, especially for your Click track. Here, you can adjust MIDI features (such as real-time properties) and access Beat Detective, which we will cover in Chapter 5.
Audio Suite Menu	The Audio Suite menu options include available Audio Suite effects and tools. Unlike RTAS effects, Audio Suite effects do not work in real time – they must process some audio and save the result. Here, you will find a complete stock of many of the same tools found when adding a plug-in to an insert on a track. We will cover effects in more detail in Chapters 3, 4, and 6.
Options menu	The Options menu options include the ability to set destructive or non-destructive recording options, loop recording, QuickPunch for recording, Video options, preroll and postroll options, and prefader metering. A very important option in this menu is the last one, Low Latency Monitoring. Low Latency Monitoring will help you in cases where your lack of processing power causes delays, echoes, loss of use of plug-ins, etc. When you toggle this setting, you will be able to fine-tune your Pro Tools system to record and operate with very little to no latency whatsoever. Large H/W Buffer Sizes in the Playback Engine will cause delay as well, so pay close attention to how you tune these two settings in tandem.
Setup menu	The Setup menu options include settings for hardware, disk allocation, peripherals, I/O, MIDI, as well as access to your Pro Tools Preferences the Playback Engine. This is one of the most widely used menu options in Pro Tools.
Window menu	The Window menu options include how to change the way you view your windows, how they are arranged, which windows you will see (such as the Mix, Edit, MIDI Editor, Score Editor, and MIDI Event List), your Workspace, and options to open troubleshooting-based tools such as the System Usage window and the Disk Space window, which are used to help you understand how your resources are being distributed and used by Pro Tools 8 LE.
Help menu	The Help menu options include reference guides, online help, a link for getting updates, and a search section.

Now that you have worked with Pro Tools menus and know how to get around, let's take a look at the Pro Tools preferences settings found within Pro Tools 8 LE.

2.12 Setting Preferences

The Setup menu within Pro Tools 8 LE leads the way to accessing Pro Tools Preferences. The Preferences dialog is seen in Figure 2.21.

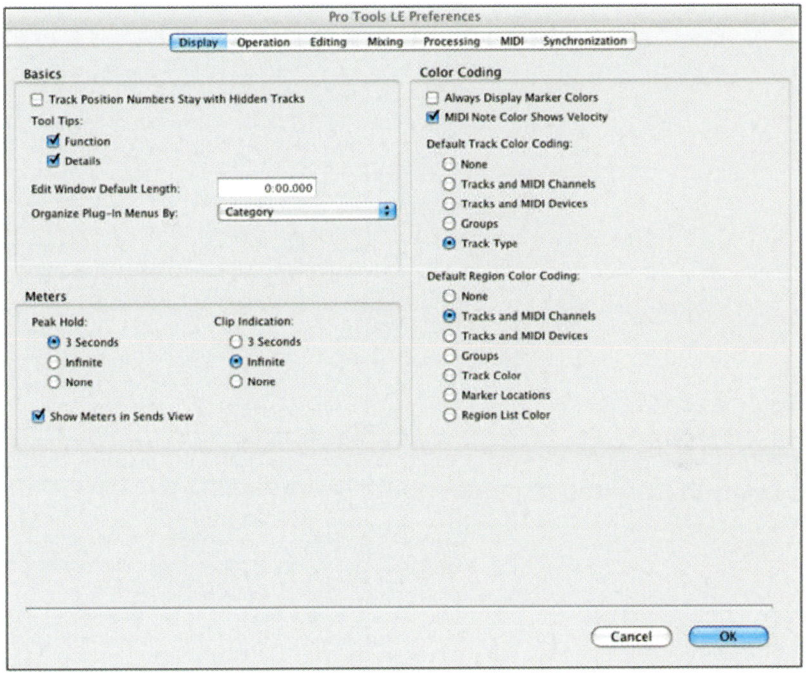

Figure 2.21 Pro Tools 8 LE Preferences.

There aren't a tremendous amount of changes in the preferences option in the new version of Pro Tools (as you can see by comparing Figs. 2.21 and 2.22), but a few do in fact make a difference. Some options have been consolidated and some new functionality has been added. The only addition in the Display settings seen in the figure is the added MIDI Note Color Shows Velocity option under Color Coding.

Other preferences you can adjust are found under the Operation tab. The Operation tab now has new options such as the Latch Forward/Rewind option, the Automatically Create New Playlists When Loop Recording option for Track Compositing (covered in Chapter 4), and the Show Quick Start dialog when Pro Tools starts option, which we configured earlier in this chapter to open a session template. Figure 2.23 shows the Operation tab in Preferences. Take note of the Auto Backup (also known as AutoSave) section that we will discuss at the end of this chapter.

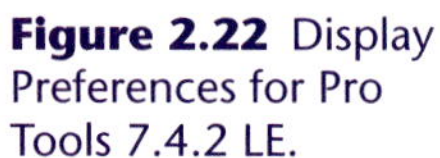

Figure 2.22 Display Preferences for Pro Tools 7.4.2 LE.

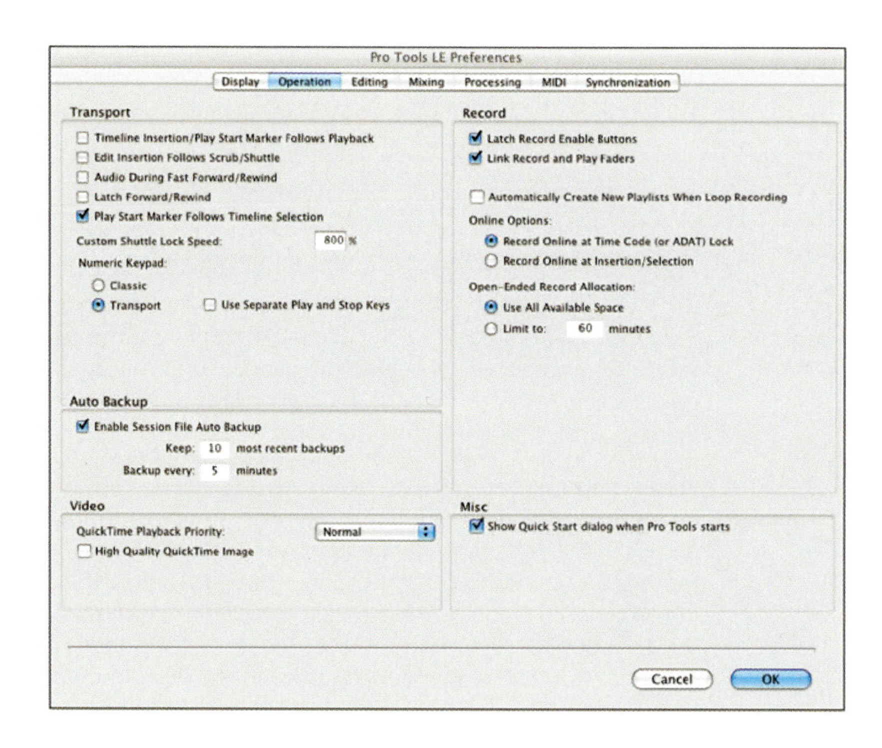

Figure 2.23 The Operation tab in Preferences.

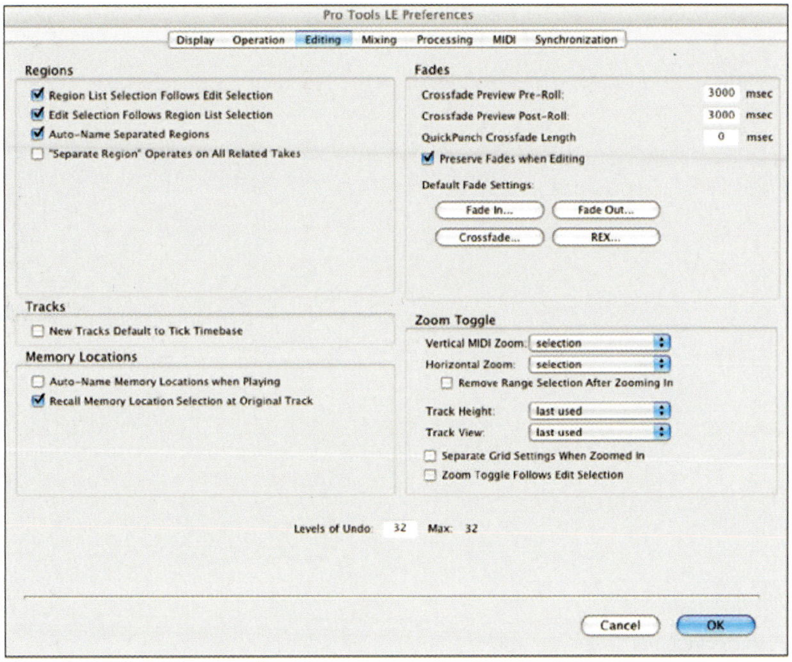

Figure 2.24 The Editing tab in Preferences.

Figure 2.24 shows the Editing options within Preferences. Here, you can work with Regions options for easier searching and manipulation.

You can also work with Zoom functions that will come in very handy when working through Chapter 6, Editing. When editing, you will use the Zoom tool (among others) to edit your recorded work. Most changes made here affect how you work with and view items in the Edit window.

Mixing preferences offer you advanced mixing options. This is where you can adjust your automation options. Most changes made here affect how you work with and view items in the Mix window. Figure 2.25 shows the Mixing tab in Preferences.

In Figure 2.26, we view the Processing tab in Preferences. Here, you can adjust AudioSuite options, where Dither is selected as a default plug-in. Dithering will be covered later in Chapter 6. You should take note of the Elastic Audio section which will come in handy when working with loops and other regions, and which can be enabled to be active when you create new tracks.

Also take note of the Import preferences. Here, you can also adjust the types of files you create and import into Pro Tools 8 LE. If you select the Convert Imported ".wav" Files to AES31/Broadcast Wave option, you enable the interchange of audio sessions and projects between dissimilar DAWs.

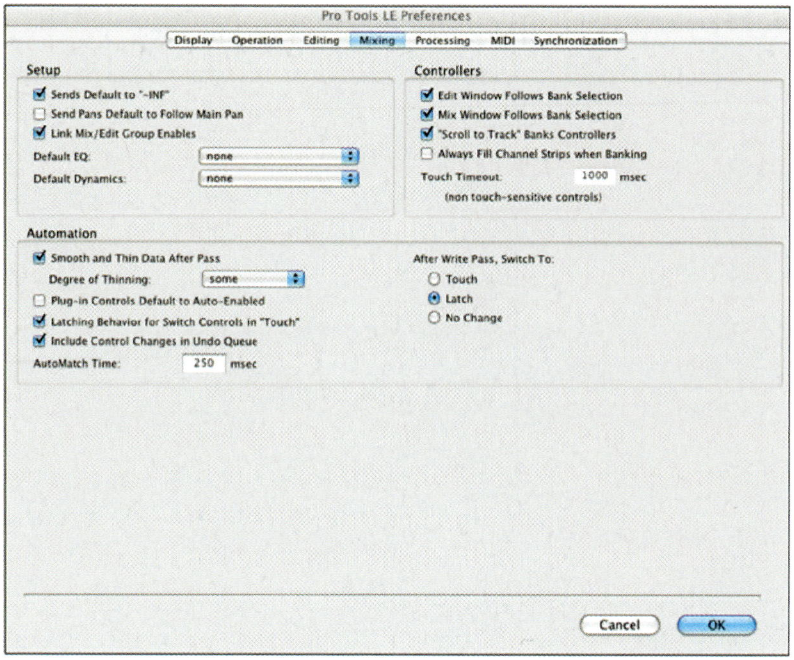

Figure 2.25 The Mixing tab in Preferences.

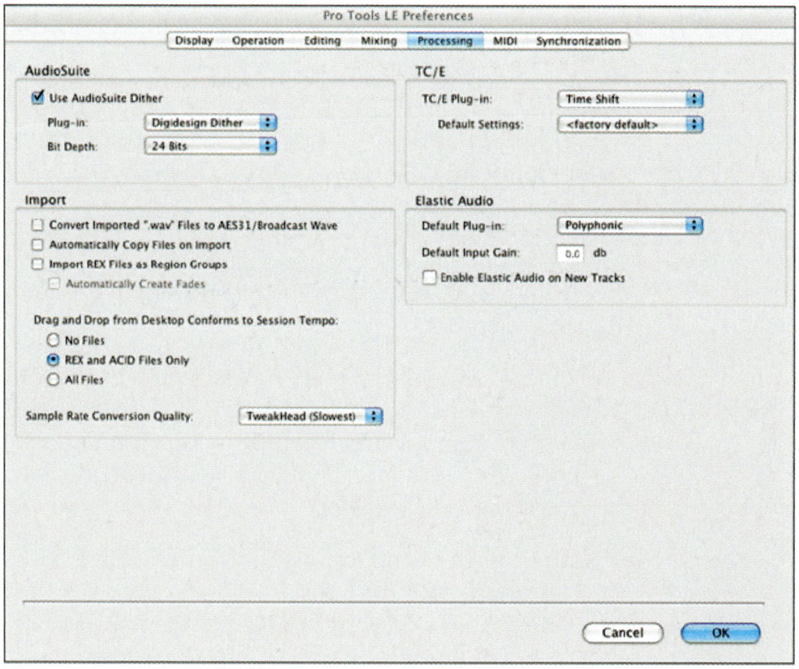

Figure 2.26 The Processing tab in Preferences.

In Figure 2.27, we find the biggest preference changes compared to older versions, the MIDI tab. Pro Tools 8 LE comes with many new MIDI-based functions, many of which are configurable here.

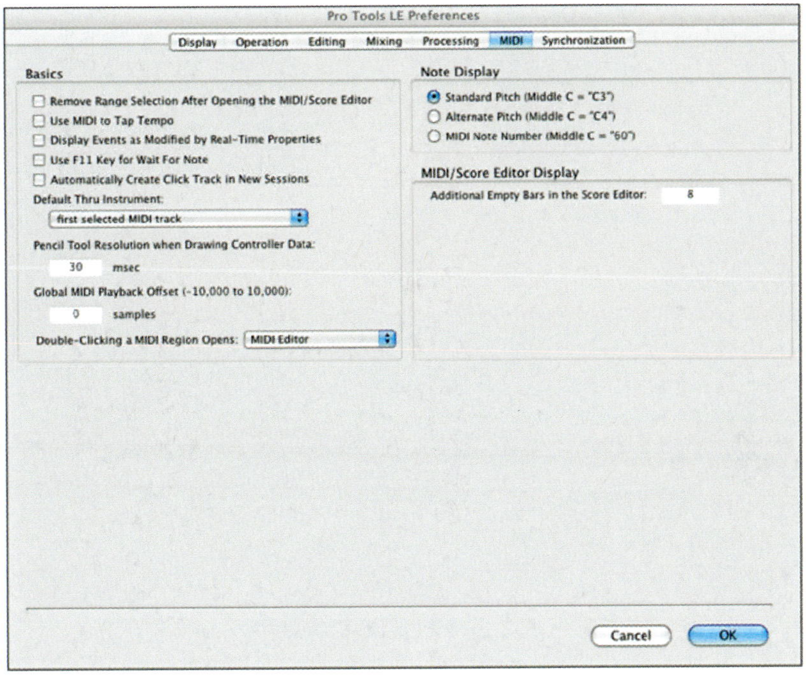

Figure 2.27 The MIDI tab in Preferences.

Some of the older Pro Tools LE 7.4.2 options are now moved around within the MIDI Editor, as well as the Score Editor windows. Options such as Play MIDI Notes When Editing have been moved to these other windows.

Finally, in Figure 2.28, you can see the Synchronization preferences. Although we already made synchronization adjustments in the Peripherals section earlier in the chapter, you can make additional adjustments here such as how Pro Tools 8 LE chases memory locations and how it follows Edit insertions.

All of these sections will be covered in more detail as we work through the rest of this chapter and book. For now, be aware that the Preferences dialog contains many options you may have to adjust while working in a production environment. Next, we will look at how to route your signal, by using sends, inserts, the bus, and more.

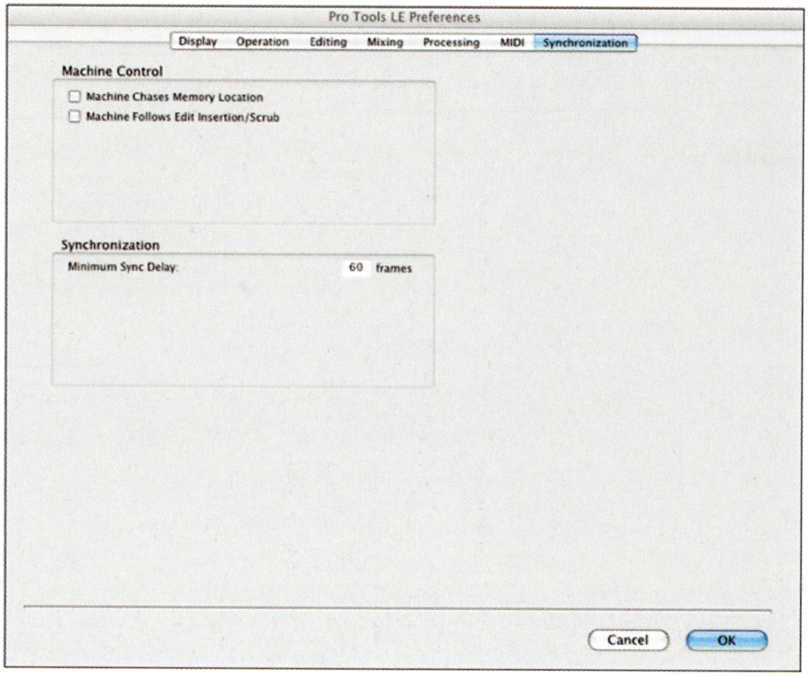

Figure 2.28 Viewing the Synchronization tab in Preferences.

2.13 Using Sends

Sends are used to route a track's signal to an additional location, such as a bus (Auxiliary Input track) that contains a reverb plug-in. Pro Tools 8 LE lets you insert up to 10 sends (A–E and F–J) on each Audio track, Auxiliary Input, or Instrument track.

Send controls can be displayed and edited from the Mix or Edit windows or in their own output windows. They can be assigned to available output and bus paths (main or subpaths) in mono, stereo, or any of the supported multichannel formats for surround mixing. Also, each send can have multiple assignments, such as sending to multiple output and bus paths. An effect must be returned to the mix through an Auxiliary Input, Audio track, or Instrument track to be audible in Pro Tools. Effects can also be monitored and processed through an Auxiliary Input (or Instrument track) recorded to audio tracks, and bounced to disk as needed.

Figure 2.29 shows how you can modify a send in the Mix window. Note that you can select an option for the bus which was configured in the I/O Setup earlier in the chapter.

Figure 2.29 Configuring a send in the Mix window.

The Send can be assigned to mono or stereo (or any of the supported multichannel formats for surround mixing) outputs or bus paths. Figure 2.30 shows the send output level adjuster, which you need to adjust if it is, for example, too low in the mix.

Set the output level of the send in the Send window by adjusting the Send Level fader. Alternatively, you adjust this send level by setting the gain on the fader to 0 (0 dB) by Alt-clicking (Vista/XP) or Option-clicking (Leopard) the Send Level fader. To remove a send from a track, click the Sends button on the track and choose No Send from the pop-up menu. With Pro Tools 8 LE, you can now use up to 10 sends per channel. You can also now drag sends, or Ctrl- or Option-click the send, to copy it over to another track as long as the tracks are matching – you cannot copy or drag a send from a stereo track to a mono track.

Figure 2.30 Adjusting the Send output level.

Tip

Sends may be designated as either prefader or postfader. Prefader means that the send or effect is inserted before the fader, thus allowing the send to be controlled independently of the channel fader. Postfader means that the send is inserted after the fader and is being controlled in proportion with the channel fader. Any changes made to the channel fader will affect the level of a postfader send.

2.14 Using inserts

Inserts operate partly like sends do. An insert effect is in series with the track's signal. A send/return effects' loop is in parallel with the track's signal. When using an insert effect, you must adjust the dry/wet mix control for the desired amount of effect. When using an effect send and return (effects loop), you set the effect to 100% wet, then adjust the send level to change the amount of effect on a track. An insert effect is usually used for compression, limiting, and gating. A send effect is usually used for reverb, echo, chorus, and flanging. In fact, using a send for compression, limiting, or gating does not work.

You can drag and copy inserts and sends in a similar manner and many of the same rules for their use apply. Pro Tools lets you use up to ten inserts on each Audio track, Auxiliary Input, Master Fader, or Instrument track. Each insert can be a software plug-in insert, a hardware insert, or an instrument plug-in. Inserts provide the ability to add many plug-ins and hardware effects, route the signal from the track through the effect of your choice, and automatically return it to the same track or send it to another track for further processing. It can operate much like an effects insert found on most amplifiers and effects processors sold today.

An insert can be either a software Digital Signal Processor (DSP) based on a plug-in or a hardware insert. A Pro Tools insert routes the signal from the track to a plug-in or external hardware effect of your choice and automatically returns it to the same track. Inserts do not alter the original audio source files but process audio in real time during playback. You can permanently apply real-time effects to tracks by recording or bouncing the effect to disk. Plug-in inserts are software inserts that process audio material on a track in real time. For example, the EQ, compressor, and delay plug-ins supplied with your Pro Tools system can be used as real-time plug-in inserts. Additional real-time plug-ins are available from Digidesign and many third-party developers.

Hardware inserts send and return the signal to the corresponding input and output channels of an audio interface, which can be connected to outboard effects. Some Instrument plug-ins (such as the Virus Indigo synth, http://www.vintagesynth.com) accept audio from the track input, letting you use them as processing plug-ins. Inserts added to Audio, Auxiliary Input, and Instrument tracks are considered prefader. Inserts on Master Faders are considered post-faders. You can cause clipping if you boost their gain to extremes, especially on tracks recorded at high amplitude. Watch on-screen metering for indication of clipping.

Inserts can be bypassed or made inactive, which help you test their viability in the mix. When more than one insert is used on a track, they are processed in series. Inserts can be used as either stand-alone processes on single tracks or you can save processing power by using them as a "shared resource." An insert can be used as a shared resource in a send-and-return arrangement by bussing signals from several tracks to an Auxiliary Input and then applying the insert to the Auxiliary Input track. By doing so, you can then configure each track to route through this Auxiliary Input, saving the time (and power) that would be used by adding the plug-in to each individual track. Both the Mix and Edit windows can be configured to show or hide inserts. Plug-in

Figure 2.31 Configuring Inserts in the Mix window.

windows provide complete access to plug-in controls. Figure 2.31 shows how you can configure Inserts for use.

Next, we cover how to use the bus to send your signal just about anywhere within your DAW!

2.15 Configuring routing

In this section, we cover how to route your signal from one place to another in Pro Tools 8 LE. Figure 2.32 shows the routing of a signal from the I/O section within the Mix window (also found within the Edit window).

To properly route the signal, you simply need to know the path and configure it.

For example, let's say you had an Audio track, an Auxiliary Input track, and a Master Fader. The path would be Stereo => Auxiliary => Master Fader. So, how would this look in Pro Tools 8 LE? You would configure your Audio Track's I/O section such that the input would be the preamplifier you have a microphone connected to and the output would be to the Auxiliary Input track instead of Analog 1–2. Then you would route the Auxiliary Output tracks to the Master Fader.

Although not the easiest thing in the world to figure out, understanding routing is actually very easy once you understand the I/O setup and how the path is actually routed through Pro Tools.

Figure 2.32 Routing the signal from location to location.

2.16 Saving your Session

Although we have much more work to do, you should know how to save your session before you start to write a passage and/or record it. Figure 2.33 shows the Save dialog used in Pro Tools 8 LE. Make sure that you save your session to a drive you prepared for Pro Tools. Earlier in Chapter 1 and in this chapter, we covered how to prepare a drive for use with Pro Tools 8 LE.

It is very important to save your session correctly so that you can open it again and use it without issue. However, make sure you not only save your session but also back it up to a secondary location if it's a very important piece of work. Chapter 7 covers backup and restoration tips but until then make sure that you save a copy of your work and have a backup for safekeeping.

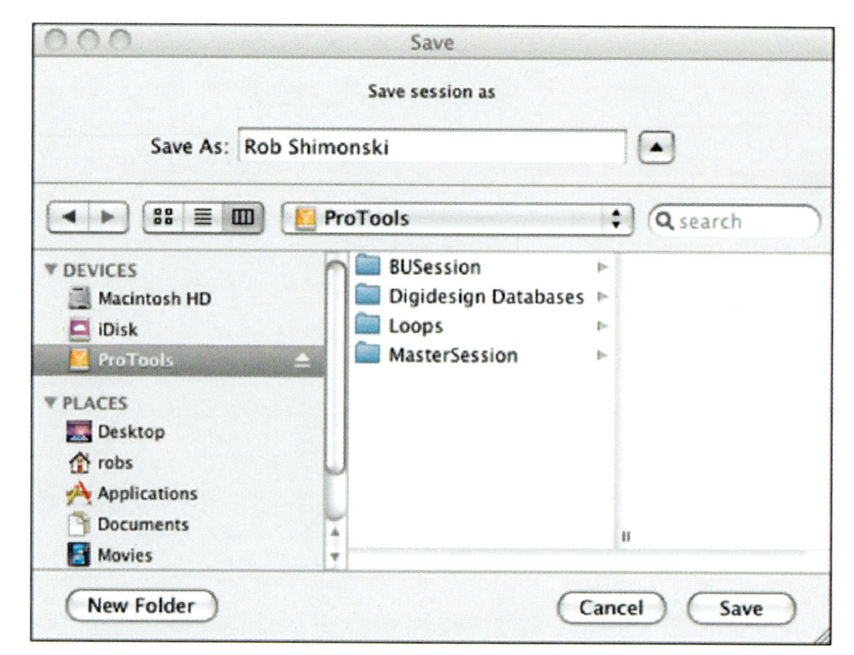

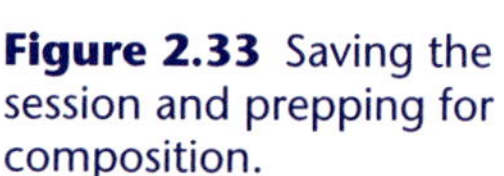
Figure 2.33 Saving the session and prepping for composition.

Using Autosave

A quick solution to backup and recovery is the Pro Tools AutoSave option. Although AutoSave is enabled by default, many engineers and producers habitually hit the save key command to quickly save their work after doing anything major. In Pro Tools Preferences, you can adjust how many saves your copy of Pro Tools will keep in its history buffers. You can use the standard defaults such as setting Keep: from the default of 10, to a higher number if you feel that more incremental backups are needed, or change the Backup Every: field to a another time setting than 5 minutes if you want session backups taken sooner and/or later than the default.

It is recommended that you change these per your working habits and needs. Chapter 7 covers other considerations you can take into account for keeping your sessions safe and available for use. AutoSave is very important, as you may experience any number of issues when working, such as power loss. It's recommended that you use an uninterruptible power supply system (UPS) on any production system in which the loss of power may ruin your system or session. When designing a studio, power is one of the main considerations you need to address. Not only do you need to know how to supply it but also how to keep it continuous so that your operation is not interrupted.

2.17 Summary

Now that you have set up your session, you are ready to produce some music! The next chapter covers composition. Within the next chapter, we will learn how to record with MIDI, compose a work, use the Score Editor, and also touch on the use of Sibelius.

In this chapter

3 Composing

Now that you are familiar with Pro Tools, let's get to work. In this chapter, we will look at how Pro Tools 8 LE allows you to score and compose like a true professional. With Pro Tools 8 LE, you can now build upon your creativity with new and enhanced tools such as the fully integrated Musical Instrument Digital Interface (MIDI) and Score Editor windows, which allow for more possibilities when editing and working with MIDI. This chapter covers scoring, composition, and the use of MIDI devices, as well as the new MIDI functionality in Pro Tools 8 LE and Sibelius.

3.1 Introduction

Now that you have a working DAW, Pro Tools 8 LE installed, and a session up and running, it's time to write some music! Whether you are producing an artist, helping engineer a musical work, or helping with the arrangement of a song, knowing how to use scoring tools will help you become more productive. In this chapter, we will cover the nuts and bolts of configuring devices for composing and working with MIDI and the Score Editor and will also cover the use of Sibelius 5.

Composing (or composition) is the process of creating an original piece of music. A score is the final product. Good composition leads to well-written songs, as well as the ability for others who have never seen or heard the work to understand what was intended through musical notation. Obviously having a background in songwriting and musical structure and notation makes it easier to compose, but whether you are a pro or a beginner, Pro Tools (and other tools such as Sibelius, which we will cover later) can help you learn and master the process.

Pro Tools in its newest form comes with the features needed to use it as a composition and scoring tool, most notably with the addition of a new tool called the Score Editor. This chapter will show you the basics of composing and scoring with Pro Tools and what tools you will need to get it done. First, we need to understand role of MIDI in composition and creating a score.

3.2 Using MIDI

The MIDI standard, which was released in 1983, has become the de facto standard for communication of musical performance data. Today, it's used not only for audio but also for lighting and other industries serving multiple purposes. Its main use, however, can be found in recording and composition work. Since MIDI data can be converted to musical notation, composing and MIDI go hand in hand. When producing, you look for efficiency, and there is nothing more efficient than sound replacement if you want flexibility. Sound replacement allows you to take any sound from your library, and swap it out for what is currently in use.

Before we cover composition, we need to prepare to produce material on a MIDI keyboard or in Pro Tools 8 LE's Edit window. You will need to create a MIDI track and enable it for recording. Once you have recorded your work, you can then edit it. That being said, if you choose to write MIDI with the MIDI Editor, you can do so with a simple Pencil tool and record and edit simultaneously. Before we learn to record and edit MIDI, we first need to learn the technical definition of MIDI and how it can be used with Pro Tools 8 LE.

MIDI is a protocol for a specified digital signal that represents a musical performance, sent across a cable from one MIDI device to another MIDI device. These devices require a MIDI cable based on specific standards. MIDI is also considered a language that computers speak, helping MIDI-enabled devices connected to it send real-time events between them for recording. MIDI communicates using 16 channels, which allows up to 16 MIDI instruments to be played from one MIDI interface. You can add a second MIDI interface and open up an additional 16 channels.

What makes MIDI so interesting (and different from the audio recording covered in the next chapter) is that MIDI does not create a waveform nor does a MIDI device transmit sound. Instead it sends musical performance data (note on, note off, key velocity, etc.) to a receiver such as a MIDI interface working with Pro Tools, Reason, Sibelius or many other MIDI recording tools.

Figure 3.1 shows an audio recording. On the track, you can see that a waveform has been captured and is ready to be edited, mixed, and bounced. The region is saved, and it can be loaded up as needed. The sound that was recorded can be edited, but it cannot be changed into a completely new sound.

Figure 3.1 Viewing a waveform.

Figure 3.2 shows a track in the MIDI Editor with notes recorded from a MIDI keyboard. As you can see, only note information is recorded. Any sound sample can be inserted in the project from a library and can be triggered by MIDI events.

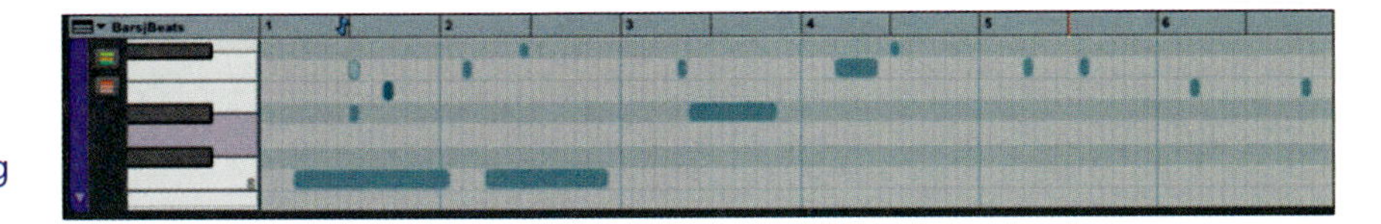

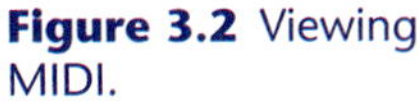

Figure 3.2 Viewing MIDI.

Once you connect your devices, you can begin to work with them in the Edit or Mix windows by setting up a MIDI or Instrument track or within the MIDI Editor. MIDI devices and instruments are connected with the help of the Audio MIDI Setup window. You can open this by going to Setup menu, then selecting MIDI and then MIDI Studio. This will open the Audio MIDI Setup and allow you to configure both audio and MIDI devices on your system. You can click on Add Device, test and scan for devices, and more. Once you have scanned for your devices, you can create and name a configuration for it, so they will be ready for use. It's recommended that when you build a new configuration, you set up and test each device in the MIDI Studio.

A MIDI Controller can be a piano-style keyboard, drum pads, or breath controller. A MIDI device can be any device that works with MIDI, such as a MIDI splitter, MIDI Controller or MIDI interface.

You do not need to use a MIDI Controller to record with, but it does help tremendously. There are many different MIDI devices out there providing a variety of options to fit within your budget and needs. An example of an affordable yet feature-rich controller is the M-Audio Keystation 61ES. It has 61 notes and controls for pitch, velocity, and modulation. It also has a foot pedal jack to control sustain and damping.

Using Reason

You can use a tool like Reason (http://www.propellerheads.se/) to act as a sequencer. A sequencer is software or hardware that records MIDI events. Many times engineers like to rewire Reason (as seen in Figure 3.3) into Pro Tools to use it within their session. This is done with the ReWire tool, which allows you to add Reason to a track within Pro Tools for real-time use and effects. This can be done by clicking on an open insert on a stereo audio track, selecting the multi-channel plug in `Reason' from within the Instrument section and you can now run Reason and Pro Tools simultaneously in real time.

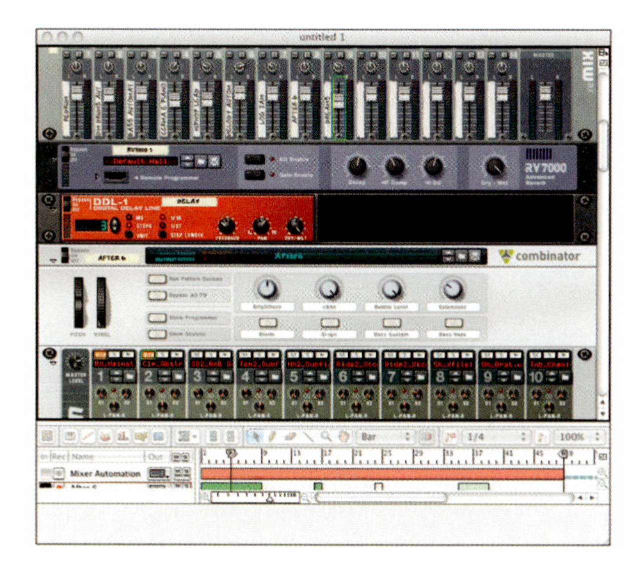

Figure 3.3 Using Reason with Pro Tools.

Ok, let's get started with recording MIDI. First, open a session. By opening a session, you can create and enable tracks. Figure 3.4 shows MIDI tracks in use within the Edit window as well as the new integrated MIDI and Score Editors built directly into the interface.

Figure 3.4 Recording and editing MIDI in the Edit window.

First, you will want to create a MIDI track to record on. As noted in Chapter 2, you can use an Instrument track to record MIDI data as well. To create a MIDI track, simply go to the Track menu and select New. Then, create the tracks you need and save the session. You can name the tracks as needed (e.g., keyboard) and move them into place.

3.3 Record MIDI Tracks

Once your tracks are created, configured, and ready, your next step is either route your signal or to "record enable" or "arm" the tracks (Fig. 3.5). Simply click on the record button in the Mix window or Edit window to record enable your tracks. Press the spacebar of the keyboard or click on the Play button in the Transport tools to start recording.

Figure 3.5 Record enabling your tracks (the red rectangles with dots are blinking).

To further configure your MIDI tracks, you should know how to manage the MIDI signal path. If you look at the Mix window in Fig. 3.5, you will see that the MIDI track's Input is set to All. That means this track will accept MIDI data on any channel from any port. This setting allows you to set up and configure multiple MIDI devices and use all of them simultaneously.

As you create new MIDI devices in MIDI Studio, you will find they show up as specific inputs ready for use. If you are recording multiple tracks at once and want to keep each performance to its own track, then you will want to specify the input of each track as a specific MIDI device. The output section identifies the device that you will send the signal to.

Be sure to check that the MIDI input device is located in the MIDI Input Enable dialog found in the Setup menu by going to MIDI and then Input Devices. Here, in the MIDI Input Enable dialog, the Mbox 2, 003, or any other controller must be enabled (selected). Once you enable the track to record and begin to play, if everything is routed correctly (and all MIDI devices are functioning correctly), you will be ready to go. Once your track is ready and you want to record, you can verify that your track is armed, then hit the record button in the Transport section of the Edit window and begin then press the spacebar or the play button in the Transport to begin.

Using the MIDI Input Filter

There will be times that you will want to record only some of the performance gestures (MIDI messages) on the MIDI network! So, how do you remove the MIDI messages that you don't want to record? You can use the MIDI Input Filter as seen in Fig. 3.6.

The MIDI Input Filter can be used to filter out any MIDI messages that you do not want recorded. You can select All, Only..., and All Except ... in the Record options, which will allow you to adjust what you do or do not want recorded. You can remove (as seen here) Mono and Polyphonic Aftertouch, Program Changes, Pitch Bend, and Notes for your channels. In the Controllers section, you can adjust which controllers are to be affected. You can also (by clicking the up and down arrows in the Controllers fields) add and/or remove a controller as seen in Fig. 3.7.

Figure 3.6 Configuring the MIDI Input Filter.

Figure 3.7 Adding and removing Filtered MIDI Controllers.

Once you finish your recording, you can click on the stop button and save your work by going to the File menu and selecting Save, or you can also use your keyboard shortcuts: Control and S keys in XP/Vista and Command and S keys in OS X Leopard. Appendix A covers Windows and Apple keyboard shortcuts in more depth. Because recording MIDI at this point is nearly identical to recording audio, we will finish the recording discussion in the next chapter.

If you want to monitor playback in your monitors or headphones or monitor MIDI tracks while recording, you may need to configure the MIDI Thru setting. You can configure this in the Options menu.

Finally, when recording and working with MIDI, take note of a new feature that makes it easier to configure your meter, tempo, and other operations. You can add a set of MIDI Controls right to the Transport area of the Edit window. You can configure this by selecting MIDI Controls using the drop-down array (Fig. 3.8).

Figure 3.8 Adding Transport features to the Edit window.

You can also add and remove synchronization functions here by selecting Synchronization. We discussed synchronization in Chapter 2 when discussing Preferences and configuring MTC. You can also select Expanded Transport to add additional control and viewing options as well as more functionality.

3.4 Editing MIDI

There are many MIDI editing tools related to composition, including the MIDI Editor and the new Score Editor. To expose the new integrated editors, simply click on the drop-down arrow found on the bottom left-hand side of the Edit window. As seen in Figure 3.9, this drop-down arrow has a line under it and the arrow faces up or down. By clicking on it, you will open or close the MIDI and Score Editors depending on its position. You must have created a MIDI or Instrument track, or you will not be given the options to use either the MIDI, or the Score Editors.

Figure 3.9 Expanding the MIDI Editor in the Edit window.

Once you have opened the MIDI Editor with the Edit window, you can switch between the MIDI and Score Editors seamlessly with the touch of a single button giving you access to all three tools within one window.

You will find there are many ways to edit MIDI in Pro Tools 8 LE. Once you have completed your recording, you can open up and use many different tools (Pencil and Grabber to name a couple) to write and edit your song. You may find that a song you wrote needs to be reworked, some notes needs to be changed, and even the velocity of some key strikes needs to be softened. Maybe the chorus and verses of your song are not in the order you wanted them, and you want to make the song shorter in overall length. Perhaps you recorded certain sections in the key of E and want to drop the key down to D or D minor. Whatever the case, this is where recording with MIDI becomes beneficial. Although you can make changes with audio waveforms, there is nothing easier than working with MIDI to make these changes. When using MIDI, changing pitch (transposing) is much easier than with its audio counterpart. Pro Tools 8 LE now uses Elastic Pitch for audio to change the pitch of a section quickly and easily. Other adjustments can be made with MIDI that cannot necessarily be done with recorded audio such as changing a rhythm by adjusting the timing codes.

You can also adjust the views before you edit MIDI, as this will change how it's edited and how you are able to manipulate it. For example, in Figure 3.10, if you click on the top left corner of the track, there is an arrow that will allow you to select what type of track you will see. Figure 3.10 shows the Region view with a recorded MIDI track.

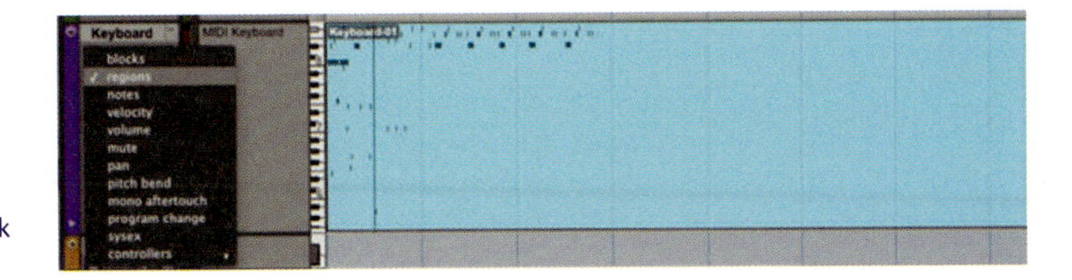

Figure 3.10 MIDI track with Region view.

To Edit MIDI in Note view, simply change the view, and you will then be able to see the work in notes. You can also adjust MIDI Controllers from this selector. Simply click on the arrow on the bottom left-hand side of the track to (+) add a track or (−) remove a track as needed.

The way you view the data is the most important thing to consider. If you know how to read musical notation, then you would want to perhaps try your hand at writing your song with the Score Editor. If you want to play and program it with MIDI, then the MIDI Editor is for you. Although you can edit the tracks you make within the Edit window with the same tools you would use to edit audio, it is recommended that you use the MIDI Editor because of how well it integrates into the Edit window and how accessible it is.

For example, record your MIDI tracks with the standard Transport tools in the Edit window. When you need to edit, drop into the MIDI Editor directly in the Edit window. Now, you can flip to the Score Editor with one button. You can also work within each Editor while using a specialized MIDI toolbar giving you access to the Pencil, Grabber, and other MIDI editing tools. The toolbar can be seen in Figure 3.11. The notation button when selected (turns blue) will turn the MIDI Editor into the Score Editor. Deselecting it turns it back to the MIDI Editor.

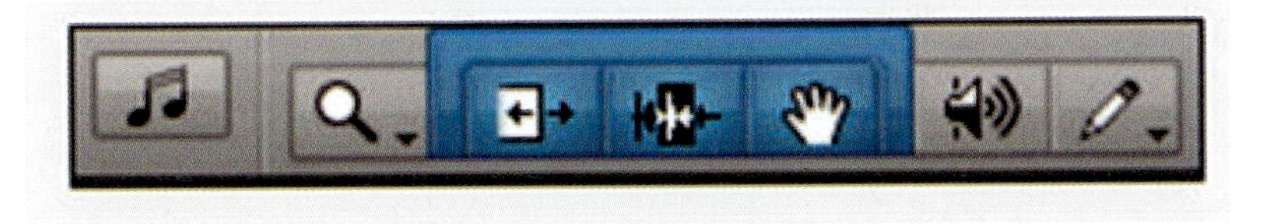
Figure 3.11 Viewing the MIDI Editor Tools.

From within this Editor, you can now transpose, work with MIDI notes, change the velocity, and score your song with the Pencil, among other things. If you open this up within the Edit window, you can record your audio, have access to regions, record and edit MIDI, and flip to the Mix window for effects and fine tuning. The toolbar provides the ability to quickly edit any sections in real time.

If you need to move or transpose MIDI notes, simply use the toolbar. Aside from the button that lets you flip to the Score Editor to compose, you will find left to right the Zoomer tool, Trimmer tool, Selector tool, Grabber (time) tool, Scrubber tool, and Pencil (Free Hand) tool.

The Zoomer tool allows you to click on MIDI timeline and shrink and/or enlarge the view. This is helpful if you need to zoom in to get closer to a small note to make a minor change. The more you zoom in, the easier the edit will be. If you hold your mouse button down when selecting the Zoomer tool, you will be given an option to change the view.

The Trimmer tool is used once MIDI data is written. As its name implies, it will allow you to click on any area within the MIDI note and drag either side to trim the note or lengthen it.

The Selector tool is used to select any MIDI note, and if you hold your click down, you can move the note to any location on the timeline.

The Grabber tool is used to grab a note and move it. You can also grab entire sections of notes by holding down your mouse button, clicking on the timeline, and highlighting the sections of notes you want to move. The Zoomer tool helps you get close to make detailed changes.

Note

MIDI notes can be transposed by dragging them up and down with the Pencil or Grabber tools.

Tip

Directly above the Trimmer, Selector, and Grabber tools is a section that will become highlighted blue when clicked (Fig. 3.11). This means that you have access to all three tools at the same time working within the timeline. The available tool will change depending on where your cursor lies within the track (top, middle, or bottom).

The next tool we will look at for editing your composition is the Pencil tool. This is the easiest tool to write MIDI data. The Pencil tool is customizable. For example, by clicking and holding your mouse button on the Pencil icon, you can change the type of Pencil tool pattern you use. You can see an example of the different types of tools you can choose from in Figure 3.12. The Pencil tool shapes are Free Hand (default), Line, Triangle, Square, Random, Parabolic, and S-Curve.

In Figure 3.12, you should also note the feature that allows you to add and subtract which subtracks you want to see, such as velocity, volume, and pitch bend.

The rest of the toolbar is also full of useful tools such as changing the Note Duration (Fig. 3.13). In the MIDI Editor, click on and hold down the mouse

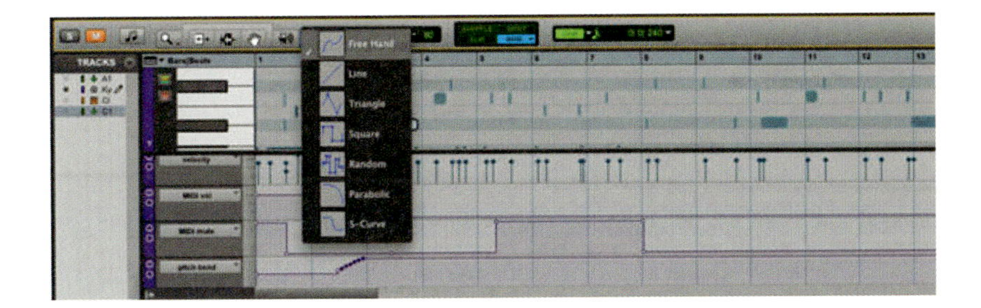

Figure 3.12 Pencil Tool menu.

button on the drop-down arrow next to the musical note to view the menu options to set which one will apply to all new notes created by default.

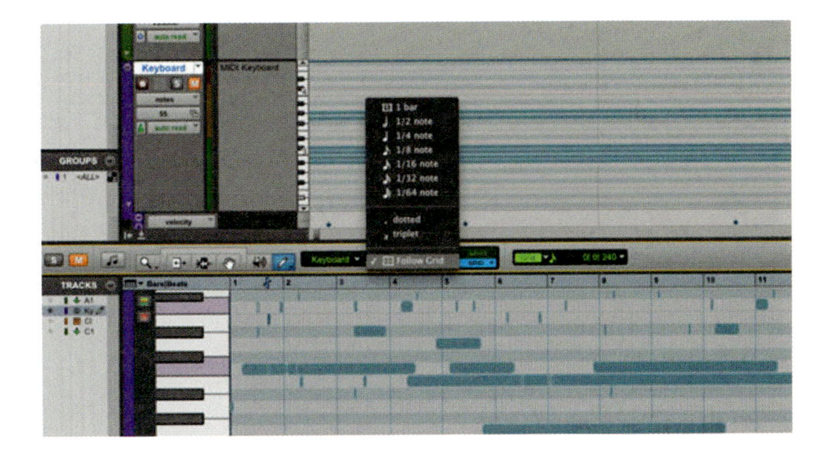

Figure 3.13 Note Duration menu.

Using the MIDI Event List

If you want to view an extremely detailed list of every MIDI event, click on the Window menu and select MIDI Event List. To customize the events, simply double-click sections of the list you wish to change. Since the list is very easy to read and extremely detailed, you can fine tune your edits here. As well, you can copy and paste events in the Event List and move them around. You can invoke a filter to hide what you do not want to see. A drop-down arrow in the top right-hand corner gives you many options you may find helpful such as Go To.... It allows you to get more views from within the list. Figure 3.14 shows the MIDI Event List.

The numbering in the tracks shows you the where the specific data is located in the session. Your Instrument tracks also appear in the MIDI Event List.

MIDI Event List

Keyboard — 18 Events

Start	Event			Length/Info
1| 3| 000	♩ F8	44	64	0| 0| 240
1| 3| 000	♩ Eb8	80	64	0| 0| 240
1| 3| 720	♩ E8	113	64	0| 0| 240
2| 1| 480	♩ F8	80	64	0| 0| 240
2| 2| 720	♩ F#8	80	64	0| 0| 240
3| 2| 240	♩ F8	80	64	0| 0| 240
3| 2| 480	♩ Eb8	80	64	0| 0| 240
4| 1| 720	♩ F8	80	64	0| 0| 240
4| 3| 000	♩ F#8	80	64	0| 0| 240
5| 2| 000	♩ F8	80	64	0| 0| 240
5| 3| 240	♩ F8	80	64	0| 0| 240
6| 1| 720	♩ E8	80	64	0| 0| 240
6| 4| 240	♩ E8	80	64	0| 0| 240
7| 2| 240	♩ F8	80	64	0| 0| 240
8| 1| 240	♩ Eb8	80	64	0| 0| 240
8| 2| 000	♩ F#8	80	64	0| 0| 240
9| 1| 480	♩ F#8	80	64	0| 0| 240
9| 3| 720	♩ Eb8	80	64	0| 0| 240

Figure 3.14 The MIDI Event List.

Another editing feature in Pro Tools is the MIDI Real-Time Properties window found within the Event menu. Unlike audio, whose pitch, duration, and tempo cannot be manipulated in real time, MIDI can be adjusted in real time. Audio can however be manipulated in real time if Elastic Audio is configured for your tracks in Pro Tools Preferences => Processing tab in the Elastic Audio section. You can use this Real-Time Properties dialog to affect your MIDI data while playing the track as well as recording (Fig. 3.15). In this dialog, you can see quantizing, duration, and velocity options.

Tip

To access Real-Time Properties in the Edit window, simply do the following. In the Edit window, in the top left section of the first track is a small drop-down arrow next to a white rectangle with lines that allows you to add I/O, Inserts, Sends, and more (Fig. 3.15).

Figure 3.15 Adding Inserts, Sends, and I/O.

As you will find, there are many, many ways to configure and use Pro Tools 8 LE for production.

In addition, you can use the Event Operations window to further edit your MIDI data. If you need to quantize your drum tracks, transpose your guitars, or flatten your performance or Select and Split Notes, the Event Operations window (Fig. 3.16) will do it. In this dialog, you can apply more edits to your work or finalize it.

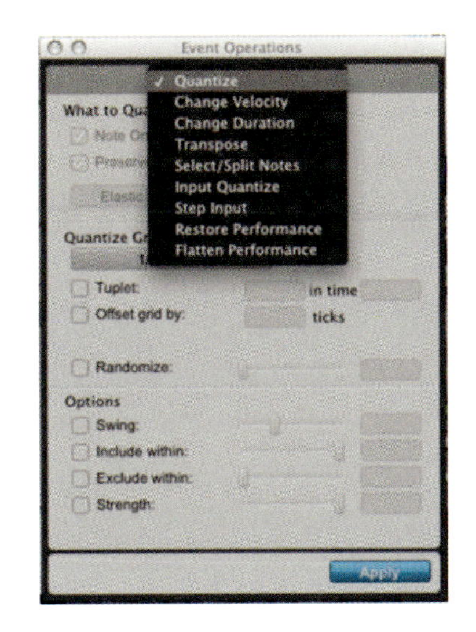

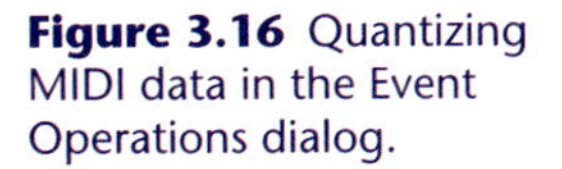

Figure 3.16 Quantizing MIDI data in the Event Operations dialog.

The Event Operations window provides a wealth of tools to help produce with MIDI. It allows you to select among the following options: Quantize, Restore Performance, Flatten Performance, Change Velocity, Change Duration, Transpose, Select/Split Notes, Input Quantize, and Step Input. Each one you select changes the subsection options below. Table 3.1 discusses the options in more detail.

Table 3.1 MIDI Operations options

Quantize	Quantize aligns MIDI notes in time with the grid of regular time intervals. You can use Beat Detective to do the same thing (when configured to use MIDI). It is used to fix note timing errors. A threshold setting allows some time flexibility within the grid.
Restore Performance	Restores your work to its last take, like an Undo function. The buffer holds up to 16 takes.
Flatten Performance	Saves your changes to the MIDI data and removes the option to restore.
Change Velocity	Set the velocity of a note to change its volume.
Change Duration	Change the length of selected notes.
Transpose	Changes the pitch of your MIDI data to a different key.
Select/Split Notes	Divides a note into two parts.
Input Quantize	Quantizes notes as you are recording, instead of after.
Step Input	Step Input is used when using a MIDI keyboard or controller to send data one note at a time, which allows you to handle faster tempos more accurately.

When working with MIDI, you can leave the Event Operations window open on the desktop. You can then select the data you want to edit, select the proper tool and settings, and click on Apply. Now that you have recorded, written, and edited MIDI, we can look more closely at the new Score Editor that comes with Pro Tools 8 LE as well as more composing features within MIDI Editor.

Note

Exporting from MIDI will be covered in Chapter 7, Importing and Delivery.

3.5 Writing with the Score and MIDI Editors

To access the Score Editor (Fig. 3.17), go to the Window menu and choose Score Editor. The Score Editor is used to compose your work into a score – you can view the musical notation that you need to manipulate! Excellent. The first thing we should do is get familiar with the editor itself and then move right into writing with it.

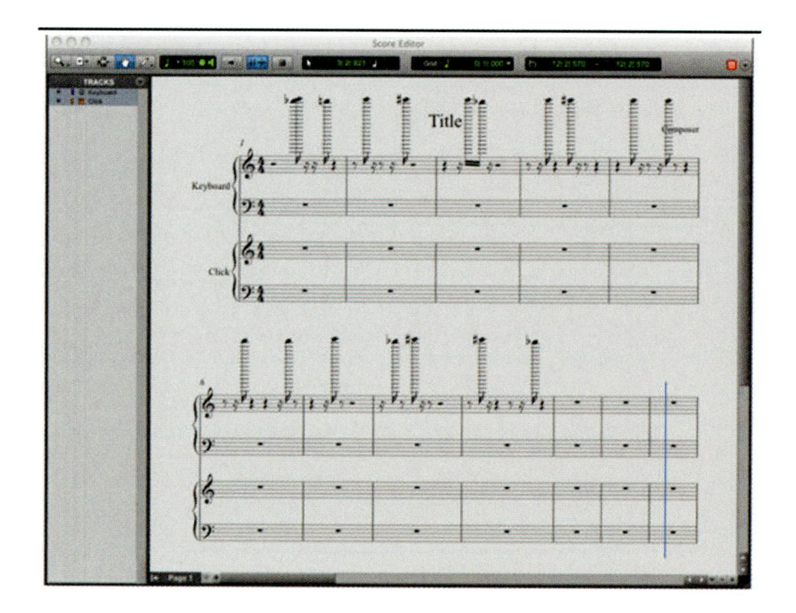

Figure 3.17 The Score Editor.

As mentioned in the previous section, you can flip back and forth between the MIDI Editor and the Score Editor if you have them opened within the Edit window. If not, then you can use the Window menu to open the Score Editor and MIDI Editor in stand-alone windows. The Score Editor is similar to the MIDI Editor. The MIDI Editor (Fig. 3.18) is used to manipulate and display MIDI data and automation and allows you to edit everything in one window.

Figure 3.18 The MIDI Editor.

Here is where understanding MIDI and configuring it becomes important. You can score without all this extra knowledge, but knowing how to use both the MIDI Editor and the Score Editor becomes extremely helpful in the production phase. Now let's write our song. First, we should set up the score. Go to the File menu and select Score Setup. In Score Setup (Fig. 3.19), you can name your song and specify the layout and what to display.

Figure 3.19 Configuring the Score.

Once you have done this, simply close the dialog and save the session. Now, when you open the Score Editor, your song title will appear as well as the other changes you made to the settings. In the same menu, there is an option to Print Score. This will allow you to print the score to either a PDF document or a printer if one is connected and configured with your DAW's computer system.

To write a song, you will need to use the MIDI Editor if you do not know how to read musical notation. If you know how to read musical notation, you can edit and write data directly into the Score Editor window. Figures 3.20 and 3.21, respectively, show the Score Editor and the MIDI Editor and how notes can be placed within them to write a song.

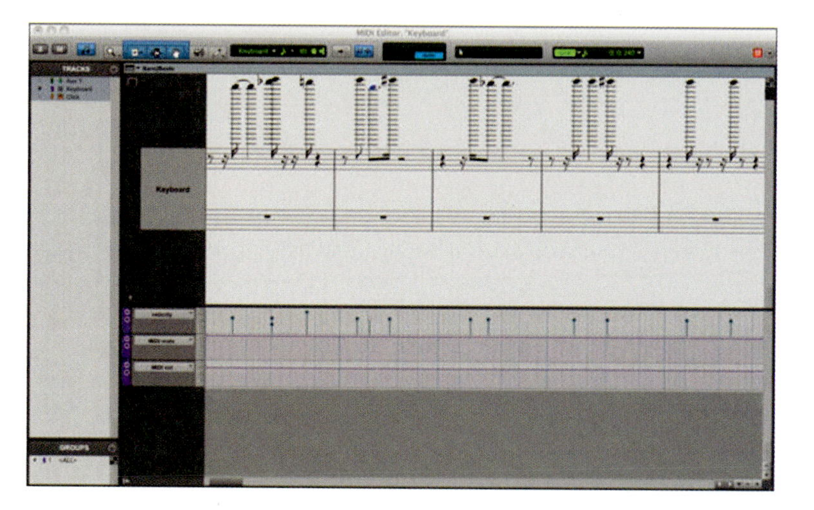

Figure 3.20 Writing in the Score Editor.

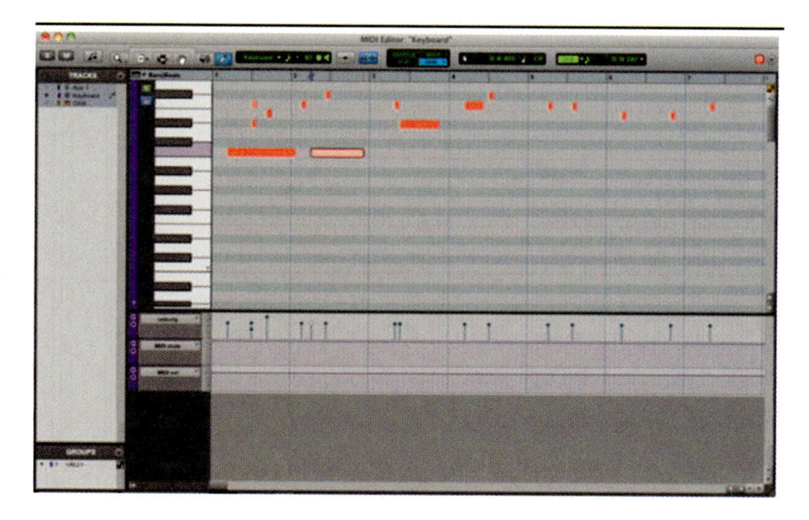

Figure 3.21 Writing in the MIDI Editor.

In the Score Editor, you can use the Pencil tool to draw on the Staff. Simply follow the same steps you used with the MIDI tools you learned and apply them to the Score Editor. Again, you can use the Notation Display enable button in the MIDI Editor to flip to the Score Editor and vice versa. Get used to using this if you are efficient in both MIDI editing and editing musical notation – you will be able to write more quickly and be more productive. As you compose, use additional subtracks to get samples, set up loops, and build your song by auditioning samples and loops for the final take. If you find that your composing is being limited and you are ready to use a musical composition and scoring tool that is 10 times as powerful as the Score Editor in Pro Tools 8 LE, you may want to write your song in MIDI with Pro Tools 8 LE, then export it to Sibelius 5 for total control over all aspects of your composition.

3.6 Using Sibelius

Although Pro Tools 8 LE is an overall superior product, there are many other products that provide more power for scoring and composition. Sibelius, Finale, and Logic provide amazing composition abilities. In Pro Tools 8 LE, you can export directly to Sibelius. For those who do not have access to Sibelius, you can register for and download a trial copy from Sibelius at www.sibelius.com. Those unfamiliar with it should download and try it. Sibelius has additional features and functionality that make it very useful for composing and scoring. Its interface is just as beefed up as Pro Tools 8 LE and so is its toolset. After working with the MIDI Editor, you will see that Sibelius (if the full version is purchased) can do much more than basic composition. For example, the product can scan in musical notation on printed documents from handwritten sources and digitize them for editing and further creativity.

To use Sibelius, you must first acquire it, install it, and configure a session for use, much like Pro Tools 8 LE. In fact, you will be very amazed at how different they look but how similar they operate, almost like distant cousins. To download a full copy or trial, you can purchase it either from online or from an authorized dealer. To install it, simply follow the instructions provided. With Sibelius, there is a set of video tutorials that walk you through every step of the way.

You can input one note at a time and use a MIDI keyboard as with Pro Tools 8 LE. You can also use the keyboard and mouse of your computer to write and edit a song. With Sibelius, you can also play your song and have Sibelius score it for you, but you will need to be able to read music and play fairly well to do it accurately. If you are not well versed on the keyboard, no problem. You will just have to take more time to enter in all the data to write the song.

Tip

You need to quantize your MIDI performance to a grid before converting it to musical notation. This is very important and will produce high-quality work when completed.

The interface (Fig. 3.22) is extremely easy to learn and use. The toolbars to the right of the screen along with your keyboard and mouse will enable you to write, edit, and test ideas quickly and easily.

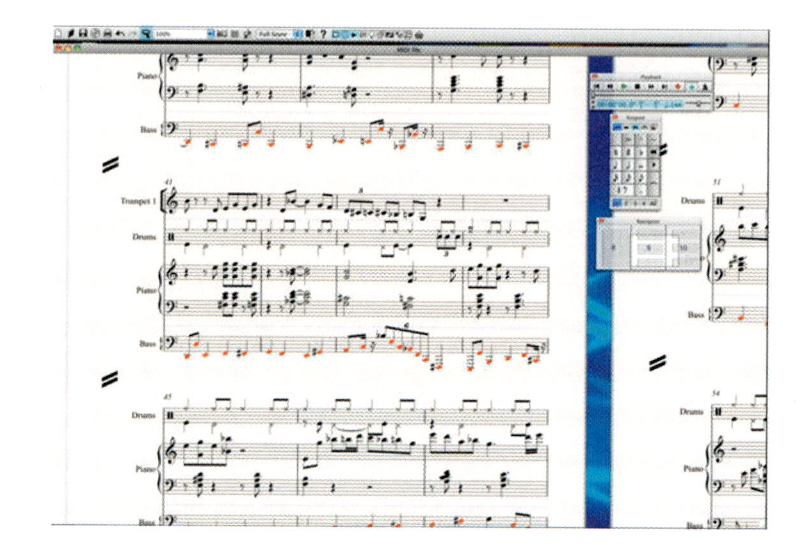

Figure 3.22 Composing with Sibelius 5.

There is also a Quick Start Wizard that will basically walk you through every step of setting up a new session and getting to work. When you first launch Sibelius, you will be greeted by this Quick Start, which can also be found in the File menu. The Quick Start dialog is seen in Figure 3.23.

Here, you can either open a recent file or start a new score. You can also open a raw MIDI file, scan in printed music, and browse an enormous amount of centralized knowledge in the teaching resources and worksheets section. You can find online resources as well as tutorials within Sibelius Help to teach you all about its use and functionality.

Sibelius also works with MIDI files. Note that MIDI data in one program is the same as MIDI data in another because MIDI is a standardized data format. As you can see in Figure 3.24, much like the Score Editor and MIDI Editor setup in Pro Tools 8 LE, you can also configure MIDI files within Sibelius to be edited as MIDI notes as well as musical notation.

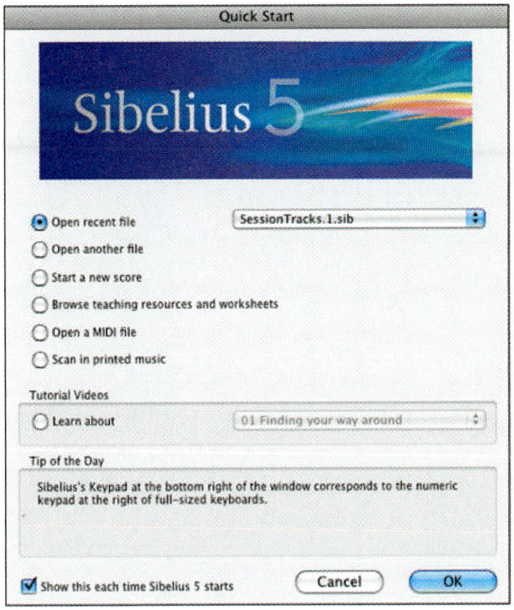

Figure 3.23 Quick Start.

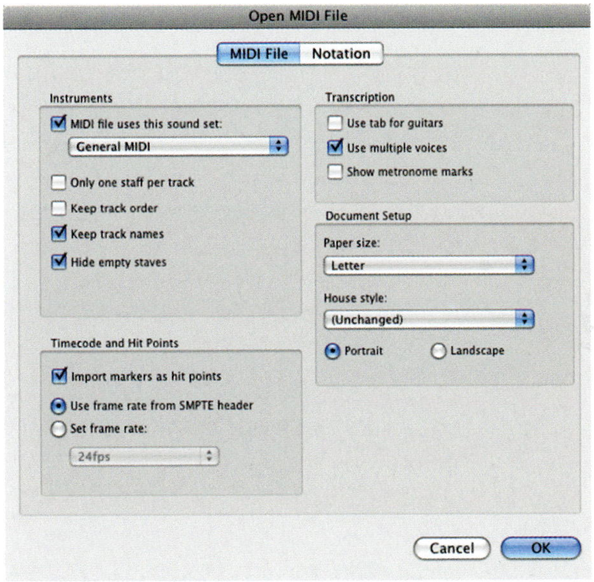

Figure 3.24 Opening a MIDI file in Sibelius 5.

Once you have created a file and opened a new session, you will be able to record, edit, and compose your work. Figure 3.25 shows the default tools used when launching a Sibelius session.

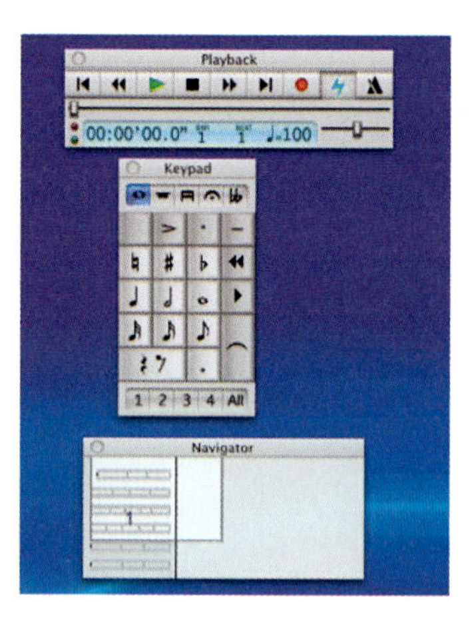

Figure 3.25 Using the Playback, Keypad, and Navigator tools.

Playback, Keypad, and Navigator will allow you to play and record your score. Playback allows you to start and stop playing the score, record to it, and make additional adjustments. The Keypad allows you to change your input so that when you work in the Staff, you can add to it directly from the Keypad. If you need something specific, click on it and then add it to your work. As you compose, you can navigate the session with Navigator. This shows you a page-by-page view of where you are in the score and allows you quick access to anywhere in the work.

To compose in Sibelius, follow the same guidelines you would when working with the Pro Tools 8 LE Score Editor – you either know how to write in MIDI or need to know how to write musical notation. As seen in Figure 3.26, you

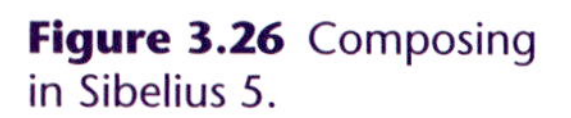

Figure 3.26 Composing in Sibelius 5.

can write music directly into Sibelius as well. You will be able to plug notes in and hear them in real time thus allowing you to create a score one note at a time.

Exporting Sibelius (*.sib) files

Note that you can save your work in Pro Tools 8 LE and then export it to Sibelius 5. With Pro Tools open, go to the Window menu and select Score Editor. With Score Editor selected, go to the Pro Tools File menu and select Send to Sibelius. You can also open the File menu, select Export, and then choose Sibelius. Pro Tools 8 LE can export a session as a MIDI file that can be opened in Sibelius as well. A Sibelius file has a *.sib file extension. Sibelius 5.x or higher is required to open *.sib files exported from Pro Tools 8 LE.

3.7 Summary

We have built a DAW, set up a session, and now recorded and edited MIDI. We learned about composing and how the new MIDI tools and features are used with the newest release of Pro Tools.

In this chapter, we learned the fundamentals of composing (or composition), which is the process of creating an original piece of music. As we mentioned, a score is the final product, and as you can see, there are many ways to create that product. Before computers, musical notation was written on paper and played from paper. Now, with everything relying so much on computers, most scoring today is done with applications such as Pro Tools 8 LE, Logic, Sibelius, and Finale.

We covered the use of the MIDI Editor and the Score Editor and how they integrate together and both appear in the main Edit window. We covered many of the tools you can use when working with MIDI. We also discussed what composition is, what tools you can use to compose with, and how to create a score.

Now that you know how to engineer and produce and/or create a song with Pro Tools 8 LE as well as Sibelius, you need to learn about waveforms, recording audio from microphones and electric instruments, and capturing live sound.

In this chapter

4 Recording

Now let's get to work recording audio, using microphones, and learning about how to track with Pro Tools 8 LE in a production environment. In this chapter, we will look at the recording process in detail, how Pro Tools 8 LE functions in a recording session, and the new functionality that enables you to be more productive when at the console. We will cover the use of recording with microphones, amplifiers, and instruments, including acoustic instruments like drums and percussion. We will also look at the numerous products that now ship with Pro Tools LE such as AIR virtual instruments.

4.1 Introduction

In Chapter 1, we covered Digital Audio Workstation (DAW) configuration and Pro Tools 8 LE preparation and installation. In Chapter 2, we started a session and got familiar with Pro Tools 8 LE by configuring the essential session setup items such as setting up the drives for recording capabilities. We configured Pro Tools for use, learned the basics of running it by learning about the different tracks you can make, studied some of the tools you can use, and discussed ways to operate within a session smoothly by using the Playback Engine. In Chapter 3, you used the system and began making tracks and learned how to work with, record and use MIDI, create and score with the MIDI and Score Editors within Pro Tools, and then move to the composition phase of the production. Here, you were able to produce the song, plan out the structure, and record any MIDI tracks needed.

Now you need to record the audio portion of your work. In this chapter, we cover the process of recording with Pro Tools 8 LE and then prepare you for the next phases – editing, mixing and mastering, and final delivery. As we learned in Chapter 2, sessions and tracks are the cornerstone of setting up and performing a recording session. When starting Pro Tools for the first time, you may not know what needs to be done to get the recording process underway. You will first need to create a session and then audio and/or MIDI tracks. Chapter 2 covered the types of tracks and how to create them.

You learned how to create Mono and Stereo Audio tracks, Instrument tracks, Auxiliary Input tracks, MIDI tracks, and a Master Fader, which is where all your tracks in a session can be routed for final processing. We also learned how to configure I/O and how to route a signal. In this chapter, we will look at using all these features to perform a recording.

Recording MIDI

In Chapter 3, we covered recording with MIDI tracks. In Chapter 1, we covered the assembly of your DAW and mentioned that to record audio, you would either need to connect up a microphone or have a direct connection. Commonly, this means that you needed cables with XLR or ¼"connectors to make an analog connection into your DAW's audio interface. MIDI Controllers needed a MIDI interface, which is built into some Pro Tools hardware.

There are differences between recording procedures for MIDI and audio. For example, MIDI and Instrument tracks can be armed to record while the session is playing. You can play your composition, record-enable a track, and work with it in real time, whereas you cannot do that with a standard Audio track. The routing of a MIDI channel also requires different I/O configurations compared with Audio tracks. In Figure 4.1, you can see some of the differences between MIDI and Audio track in the Edit window and also see a MIDI track set to record in the Mix window.

For more on recording with MIDI, please review Chapter 3, Composing.

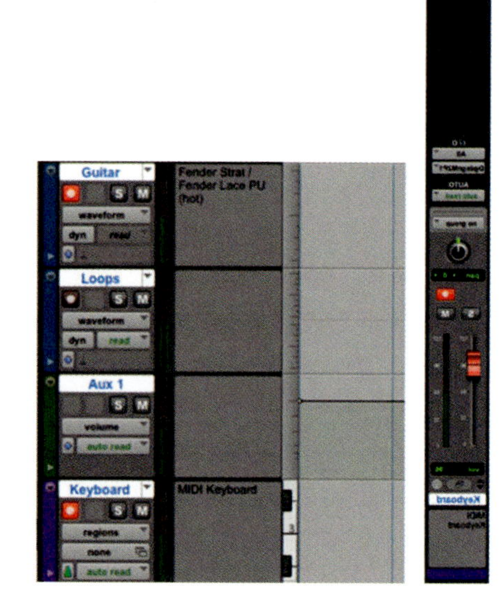

Figure 4.1 Recording with MIDI.

The process of recording is an art as well as a science. Although there are many standard techniques and methods, no two recordings are done the exact same way. Developed through research, field testing, and experimentation, certain techniques have proven to be preferred and essential for good sound. These guidelines provide a reference which you can follow to achieve the best quality for your recording. Experimenting within these guidelines is what makes recording an art as well as a science.

Successful producers and engineers bring with them talent, experience, and an extensive, well-rounded understanding of audio engineering. They have an "ear" for not only the material and its performance but for the details of the quality of the recording. Ultimately, the engineer must make decisions based on what does or what does not sound good for the recording. Producers and engineers have many different tools and use countless techniques to produce great sounding recordings. They have methods that work best for them to effectively and efficiently deliver a great sounding product. Technology itself has created a completely new way of doing professional recording in any recording studio worldwide. The DAW has truly revolutionized the recording process. Pro Audio technology fused with the computer workstation created a completely new way to create and record. Engineers have more powerful tools than ever before.

Many engineers like to bend the rules to find new and unique sounds. Maybe this is why you have decided to read this book, to learn new ways to push your system to that edge, or to be unique and strive above the rest. To create a mix so great that it's nearly flawless other than any intentional flaws, you inject into it. Finding that perfect mix or that mastering technique that makes for the perfect recording takes nothing but time, effort, and a basic understanding of the fundamentals. Consider all the priceless recordings that have already been captured and loved by millions. Each era hands down yet another set of recordings that mark their place in time. This process will always be with us as long as there is a love for music and other recorded forms of media such as movie soundtracks, television commercials, and radio broadcasts. To be able to record and create media is why you are inevitably reading a book of this kind – to learn the best possible way to capture that dream and reproduce it with the best quality possible.

So, you are ready to record. First, check your DAW configuration before we continue. As we covered in Chapter 1, preparation is the key to successful DAW management. You will save time and money and will extend the life of your investments. When working with microphones, cabling, and so on, the components are extremely sensitive to not only touch but can also basically be rendered useless if simply dropped. For example, most tube-based amplifiers contain bulb-based glass components that will shatter if dropped. Microphones are also extremely sensitive to harsh temperature changes and humidity.

Having a climate-controlled work place is important, as heat tends to build up when using multiple devices with many power sources as well as a computer system and viewing screens. Make sure that your workspace is clean and clear of obstructions and debris and that you have powered your DAW on in the proper sequence. Once you have powered your DAW on and tested it, it's time to get started with a recording session. In the next section, we will cover session preparation and configuration. We will cover the creating of tracks, routing I/O, and prepping to hit the record button to start recording a composition.

4.2 Session preparation

As we already learned in Chapter 1, the hardware we choose will determine our ability and flexibility in recording. As mentioned, the Mbox2 Micro does not allow you to record, while the Mbox2 Mini and above does. The 003 Rack has four inputs with preamps to which you can connect instruments or microphones and four additional line-level inputs with switchable gain. The Mbox2 has two inputs with preamps. The number of line inputs dictates how many line-level signals you can record at one time. You have to plan out how you will run your microphones and to which channels; this is why a patch panel bay was recommended earlier in the book when discussing DAW deployment. You may want to quickly change the preamplifier of a microphone to a different one to check for different sounds. A patch bay can facilitate making this change.

Suppose you want to record an entire drum set within minutes. Place two mics in stereo over the kit: place one mic near the snare drum and another in the kick drum. Run all the mics' balanced cables into the 003 Rack XLR mic inputs. If your drum set is in another room, you will need very long cables, a microphone snake, or a wall-mounted patch bay that is wired to your audio interface.

You can connect mics to audio interfaces with cable snakes such as those seen in Figure 4.2, which shows an EWI System patch panel (snake and stage box) that you can install in your walls or ceilings (not over ballasts!) or however you decide. It provides a way to connect over a dozen mics at once (if your audio interface has enough mic inputs).

Mobile Snakes can be used with your DAW as well if you do wish not to install a permanent cable system through your walls. You can roll it out once and use it in the studio, or you can roll it up and take it out on the road to record a band live. Also covered in Chapter 1, external outboard gear and preamplifiers allow you to expand using Optical inputs. With additional preamplifiers and mic cables, all you need now are microphones and enough interface inputs to complete the experience!

Figure 4.2 Recording Snake.

Do not forget the microphone stands and microphone clips. These are important to the process, and you will need to configure them, so you can precisely place your microphones. Microphone placement is critical to getting accurate sounds. Figure 4.3 shows a typical boom stand that can be manipulated in many ways to provide for precise microphone placement.

Figure 4.3 Microphone boom stand.

Tip

You will also find that many microphones have their own adjustment settings to assist with placement directly on the microphone casing. For example, there are low cut and roll off settings that allow you to place a mic closer to a louder sound source. Refer to each individual microphone's documentation for proper settings.

As you may recall from Chapter 1, a large part of DAW configuration when recording audio is selecting the proper microphones, preamps, and speaker cabinets to achieve the sounds you wish to hear. Live instruments, cabling, and a lot of other essentials such as batteries must also be included in the preparation before recording. Instruments tuned and maintained properly will produce the cleanest sounds and that's what you want as a recording engineer – a great sound from the source. Overprocessing your work tends to add distortion and reduce the quality of the original take. So as you will learn in the next chapter, Editing, it's wise to get what you want without too much tampering.

Note

Batteries can be easily forgotten and are sometimes the weakest link in a recording. Fresh batteries, new strings, and new skins on drum heads produce the best sounds. Batteries are needed, for example, to run mini-preamplifiers in instruments. Some condenser microphones can be powered by an internal battery as well as phantom power from the audio interface. Any instrument that uses an active preamplifier that requires power via a battery will ultimately sound better with a fresh battery when recording.

Another variable you should consider is the quality and type of instrument that you are recording. In one session, you may be recording a piano and in another an electric guitar. Each individual instrument has its own unique sound and may require different recording techniques. However ultimately, the performance of the artist will dictate the success of the recording. If the artists (or band) are not prepared, that will be audible in the recording.

Acoustically isolating your studio (as covered in Chapter 1) is also very important – especially during the recording process. For example, you may be recording an entire band, an orchestra, or a singer in a vocal booth. Each type of recording has its own needs, but what is common about these types of recordings is that each requires more room if multiple instruments or voices will be recorded at once. Busy recording studios could have multiple artists, so you may need to isolate yourself acoustically by sound proofing the location where the performance will be recorded. Keeping unwanted noise out of the final product is an absolute must when recording.

In every recording, the sound of the room plays a very important role. Microphones pick up sound reflections in the room when recording, and often they may be desired. When you listen in the control room, room reflections interfere with the sound of the speakers, thus producing a less accurate image of the material being recorded. Properly treating your studio or control room will ensure that your recording sounds good when played on many different sound systems and will allow you to efficiently create great recordings.

Note

Soundproofing and acoustical treatment are two different things. Soundproofing is isolating unwanted noises from outside the studio by using double-wall construction or cement block construction. Acoustical treatment is adding absorptive materials such as dense fiberglass panels to the room surfaces to absorb sound inside the room.

Caution

Setup is crucial to recording a project. If possible, test your equipment prior to a recording session to make sure that everything is in working order and connected correctly. Make sure that your DAW is configured and connected properly. An incorrect setting or connection may not show up as a problem until you start recording and troubleshooting while the artist is trying to record is not ideal.

Tip

Before you begin to use Pro Tools in the recording studio, make sure that you have done everything you can to preserve the sound you are capturing. As noted, treating the rooms for noise reduction and isolation and placing the microphones correctly are critical to capturing the intended sound. When you are in a sterile isolated environment free from exterior noise, you can place and cable all the microphones into your Pro Tools system and use them knowing that you will capture exactly what you intend to capture. You can place the microphones in many configurations, and it is important to choose the proper type of microphone, as that can sometimes make all the difference when recording. Chapter 1 covered the different types of microphones you can use and when you should use each type.

Sometimes a microphone can be placed in a different location to achieve a new result – remember, break the rules as you learn them! Recording is also a science, so it's important to understand fundamentals but feel free to attempt new things to keep the recording experience alive, and as spontaneous as it should be. Once the microphones are placed effectively, it is wise to get moderate recording levels from them so that you do not suffer from a type of distortion known as clipping, which can ruin the overall quality of the recorded track.

Some instruments may dictate the need to go direct into the board. Going direct means that you are bypassing the microphone and connecting the instrument straight to the audio interface. Some instruments that do not require the use of microphones to record them, such as keyboards or drum machines, would be connected to your interface with a Direct Box also known as a DI. A DI converts an unbalanced signal, such as that of a guitar or bass, to a balanced signal that is suitable for the connections of your mixing console or audio interface. In live sound situations, eliminating the microphone is beneficial, as it can reduce leakage from other instruments if acoustical isolation is not possible. However, some recordings require (or benefit from) that leakage to create the recording they desire. Live recordings are a perfect example – they sound "live" because each microphone is picking up everything in the room and creating (or enhancing) the sound you hear. This can give a recording ambiance.

Once you have a way to monitor your sound via playback through studio speaker systems (monitors) or through headphones, you are now ready to begin the recording session. Do a sample recording and playback and see if you have the levels set correctly (reaching −6 dBFS maximum in peak meter mode).

Remember that it's better to obtain a good sound on your instrument before recording than to fix it after you start. Whether tuning your drums or getting the right effect for your guitar amplifier, you will want to get the sound close to how it would sound in a finished mix. Through every step of the recording process, consider how the finished mix should sound and work toward this goal. There are many tools that you can use to enhance a recording, but getting it as close as possible before recording will save time and money in the long run.

Now, you get to open your session and prepare Pro Tools 8 LE to record sound from your configured microphones or other inputs. When a track is created, it is displayed in both the Mix and the Edit windows. In this chapter, our focus is on mastering primarily the Edit window although the Mix window is covered as well – they both perform the same functions but represent them in different ways and give you access to different tool sets. Next, we will need to create the tracks that will map to the correct preamplifiers and ultimately to the microphones.

Tip

Remember in Chapter 1 we set up our hard disk – if it is not set up as a record-enabled volume, you will not be able to record to it. If you get this error, then revisit Chapter 1 to learn how to prepare a drive for use with Pro Tools 8 LE.

4.3 Creating Tracks

Now your session is open and you decide to record a guitar track with a guitar head, cabinet, and ribbon microphone. You want to create the track, route the signal, and prep it to record. You also want to be able to hear it when you record it as well as when you play it back. Part of our song requires the use of different guitar sounds for layering. Layer is used to "fatten" your overall sound.

An example of this process is seen in Figure 4.4, where a second guitar track is added to the song.

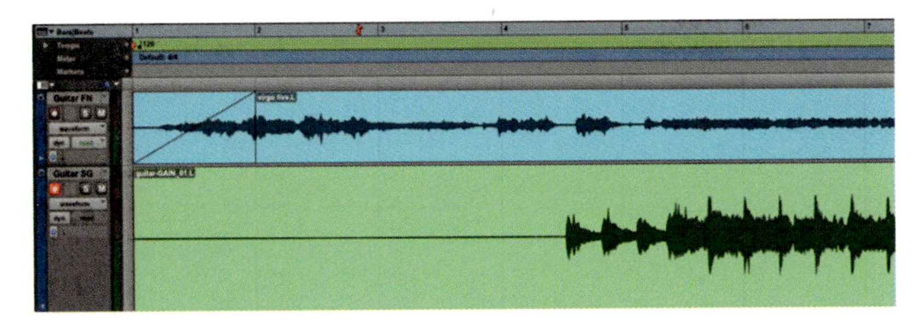

Figure 4.4 Recording to an Audio track in the Edit window.

As a refresher, a track is where you will record each instrument and vocal. In the Mix window, it is viewed as a channel strip, similar to that of a mixing console (as we noted before, channel, track, and voice are sometimes used interchangeably).

Caution

A track and a channel strip are not the same thing. A track shows the waveform. A channel strip shows the controls that affect the level and sound on a track. You need to create one channel strip per track that you are going to record. In other words, creating two tracks automatically creates two channel strips.

So, if you are going to record a singer's vocals and an acoustic guitar, at bare minimum, you would need at least two tracks. In the Mix window, you would get two tracks assigned: one for the vocals and one for the acoustic guitar. Although we cover the Edit window in Chapter 5, be aware that when you make a track, the track appears in both the Mix and Edit windows simultaneously and you can work in either window to perform your recording. The track and the channel strip appear in the Edit window and Mix window, respectively. A good rule of thumb to follow when choosing a window is that you generally cannot edit in the Mix window, but you can mix in the Edit window. Many engineers work only in the Edit window. The Edit window (as you seen in Chapter 3, Composing) is also very flexible and incorporates many new windows within it.

Next, you may want to customize your track, so you can see it better by adjusting the track's view. In the Edit window, you can click on the arrow located in the top left-hand side corner of the track you wish to adjust. This can help you see if you are clipping, for example, if you cannot see it from the default view. In Figure 4.5, you can see an example of the options available when selecting a different view. Fit to window expands the track to encompass the entire Edit window.

Figure 4.5 Adjusting how the track is viewed in the Edit window.

Other options available with Pro Tools 8 LE include the ability to create subtracks with enhanced views. You can also select the main track to be any subtrack as well (Fig. 4.6). Pro Tools is customizable so that you can work efficiently, and this new enhancement is extremely helpful, as it can, for example, eliminate the need to create new Auxiliary Input tracks to maintain volume adjustment.

You can still record with Auxiliary inputs; remember that Auxiliary Inputs can be used to bring external audio sources such as instruments and MIDI devices into a Pro Tools mix. Multiple microphones may also be mixed and sent to a single

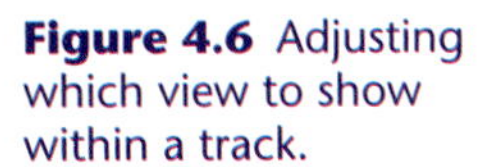
Figure 4.6 Adjusting which view to show within a track.

audio track for recording. Remember that when recording through an Auxiliary Input track, plug-ins on that channel will be printed to the Audio track.

One more very important customization you may want to make now and get used to is the enhanced views within the Edit window. For example, if you click on the drop-down arrow on the top left-hand side of the Edit window directly above the first track seen in the window, you will open the menu that will allow you to add comments section, instrument track controls, 10 inserts, 10 sends, I/O configuration, Real-Time Properties controls for MIDI, and an option to turn off the track colors (Fig. 4.7).

To get a track ready to record, record-enable it with the Record Enable button. With one or more tracks record enabled, click the Record and Play buttons in

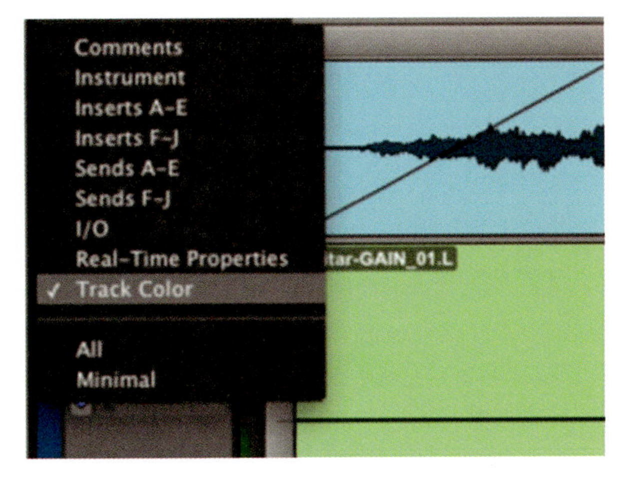

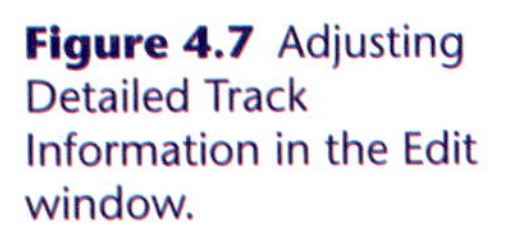
Figure 4.7 Adjusting Detailed Track Information in the Edit window.

the Transport or Edit window to start recording. Pro Tools provides a Record Safe mode that prevents tracks from being record enabled. Use Record Safe mode to protect important track recordings. To put an audio (or MIDI) track in Record Safe mode, hold down the Command key (Apple) on the keyboard and click on the track's Record Enable button. It will gray out.

Tip

Get used to using the keyboard and mouse to configure options within Pro Tools 8 LE. For example, if you need to arm multiple tracks all at once, you can hold down the Option key on your Macintosh keyboard (or the Alt key in Vista/XP), and if you record enable one track, it will record enable all your tracks. When you want to disarm all your tracks at once, simply hold down the Option key again, select one track, and you will disarm them all. Pro Tools also provides the option to solo-safe a track. This prevents the track from being muted even if you solo other tracks. This feature is useful for tracks such as Auxiliary Input tracks that are being used as a sub-mix of audio tracks, or effects busses, allowing the audio or effects track to remain in a mix even when other tracks are soloed. It is also useful to solo-safe MIDI tracks so that their playback is not affected when you solo audio tracks.

To solo-safe a track, you need to hold down the Command key and click on the Solo button of the track. This prevents the track from being muted even if you solo other tracks. The Solo button changes to a transparent color in Solo Safe mode. To go back to normal, hold down the Command key and click on it again.

Other adjustments you can make to the Edit window include how the window scrolls with you while you are recording. For example, while recording, if you do not want the cursor to follow and thus cause scrolling, you can select options such as No Scrolling After Playback (for playback and not recording purposes) and Page, moving it one page at a time from the Options menu. When you open the Options menu, select Edit Window Scrolling and select which option you want (No Scrolling, After Playback and Page).

4.4 Configure routing

Now that your microphones, instruments, and other input devices (such as MIDI devices) are placed and connected to your audio interface, they must be properly routed to each track in your Pro Tools session. Routing is essential to getting the sounds you want on each track you record.

Once tracks have been created, set the input of each track to the desired bus or interface channel the source is connected to. As an example, if you have a microphone connected to channel 1 of an interface (such as an Mbox2), you can then go to the I/O section of your Edit or Mix window and select Interface and then select the channel such as Mic/Line1. Each input follows in numerical sequence, as they are labeled on the unit itself.

The drop-down menu may appear differently depending on the type of hardware connected to your system. The type of device in use and the hardware connected to your system will dictate what interfaces you have available. As we mentioned throughout the book, your DAW's initial configuration is going to dictate the options you have available to you when working with Pro Tools 8 LE.

If you expand your view in the Edit window, you can route directly within it by choosing to view your inserts and sends. To add an insert to a track, you may need to click on the Edit window View Selector – the white rectangle with lines at the top left of the tracks – then click on a dark rectangle under Inserts and choose "plug-in." In Figure 4.8, you can see an example of adding a new plug-in to a guitar track. You can remove it just as easily by selecting "no insert" from the same menu. You can also use the Insert sections A-E or F-J to route to other parts of the I/O configuration you set up while preparing for your session.

Figure 4.8 Adding a plug-in to an Audio track in the Edit window.

Tip

Track management is not only based on colors but also on comments and proper titles. You will find that it's very important to label tracks properly within a session, as they will become unwieldy if you do not.

You can add as many as 10 plug-ins onto a track. Take note that effects process in the order in which they are set in succession from A to J. An example of multiple plug-ins installed as inserts on both guitar tracks can be seen in Figure 4.9.

Figure 4.9 Adding up to 10 effects inserts.

Now, you have to connect your instrument to your Mbox2 or 003. In Figure 4.10, you see an example of configuring the path to use "In 1 (Mono)," which is where you will plug-in your 1/4" cable.

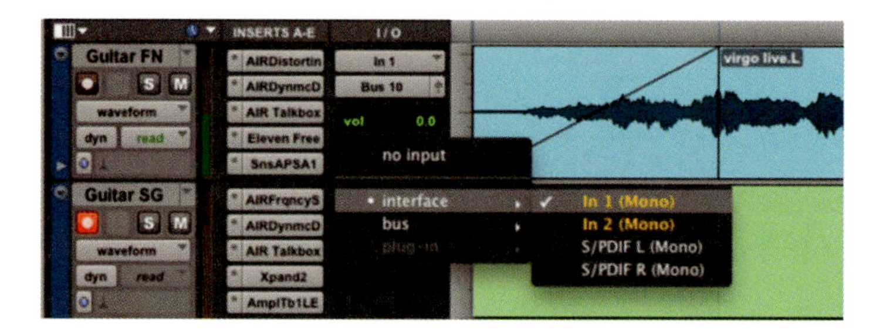

Figure 4.10 Configuring the path where Pro Tools will receive its signal.

Note

Always make sure your instruments and amplifiers are turned down completely or off so that you do not damage your hearing or the speakers, microphones, or anything else – high volumes can hurt.

To connect a guitar or other instrument directly, you will also need to configure the Mbox2 or 003 as a DI input. By selecting the DI on your hardware audio interface (Mbox2 or 003), you can now plug an instrument cable in and get to work without the use of a microphone.

Note

Again, make sure you check your batteries and cables, and change your strings or any other tangible item that may ruin your sound. Crackling, which is commonly heard when recording, is usually from damaged cables, jacks, and instrument pots. Instrument pots are other circuitry in the guitar preamplifier itself that can cause unwanted sounds in your recording. Make sure you use quality instruments and take care of them. It's recommended that for recording purposes, you have your instrument looked at by a professional to ensure it's in top shape for the recording process.

4.5 Recording with plug-ins

Digital plug-ins can be easy to use and a cost effective solution to getting many different sounds in one bundle of software. They also make recording a guitar, bass and other instruments a snap. However, they should not completely replace the use of amplifiers and speaker cabinets recorded with microphones. As we have mentioned earlier in the book ambiance recorded through microphones cannot be easily created in the digital realm without using reverb, delay and many other effects that may or may not reproduce the sound you are looking for. We also mentioned earlier in the chapter that it's important to get the sounds you want early on in the recording process to reduce the amount of work you need to do in the editing and mixing process of the workflow.

Plug-ins are nothing more than internal software applications that emulate almost any external hardware effect ever created. You can think of a plug-in as an additional module that varies the signal in the path by adding, for example, distortion or compression to the track, or by making different frequencies more noticeable via EQ. You can add delay, reverb, and countless other effects. There really is no limit to what you can find these days, and it's almost guaranteed that if it can be dreamt up, it will eventually find its way to the market.

To add a plug-in to a track, click the double arrow button on the Inserts view. From the drop-down menu, choose the desired plug-in. Plug-ins are processed in series from top to bottom in the Mix window. They are also processed from top to bottom and from left to right in the Edit window. Plug-ins will be covered in further detail in a later chapter.

As a side note, there is something to be said about vintage equipment and the sounds that they can create. You should ultimately have on hand tons of internal and external effects for whatever occasion you can think of. You may have a request from a guitarist to provide a sound that only a certain pedal effect can provide. Make sure you work with all different kinds of effects and use whatever you (and the artist) feel appropriate to build on the sound you are creating.

Tip ▼

Although you can record emulated Instrument tracks, it's not recommended that you solely rely on these for your final project. You should always use a mixture of virtual (as well as real) instruments when recording your work.

In Chapter 6, Mixing, we will cover the use of the Mix window and how to use specific plug-ins to get a great mix. Here, you will need to use emulator plug-ins in case you do not have a guitar amp and microphone. Everything you need is included with Pro Tools 8 LE, your Mbox2 or 003, and an instrument with an appropriate patch cable. We will discuss each in detail momentarily.

First, let's go over some of the tools that you can use with Pro Tools 8 LE. With this version, you get many new tools that older users of Pro Tools will really enjoy, such as the Sans Amp, which is pretty much a studio standard these days. When you install Pro Tools 8 LE (chapter 1), you are provided with many digital plug-ins such as the Digidesign AIR series of effects. Also included is a version of Eleven called Eleven Free. Digidesign did a lot of modeling work with its simulators, and when you work with them, you will agree that they do sound amazing.

As an example of adding and trying an effect, let's review Eleven Free (Fig. 4.11). Although limited to a few sounds to choose from, you can get started, see if you like what you hear and order more sounds as needed or use other plug-ins such as AmpliTube.

Figure 4.11 Using Eleven Free to record a guitar track.

With Eleven Free, you can set up a clean or distorted sound to record. If you have a very small studio setup, or just need to record an idea or two, you can get by with this. Any plug-in you choose to use will contain many features within it to help the recording process, such as the tools found on the top of the plug-in's dialog.

For example, you can adjust routing, change the factory default sound (and select between about 30 others), and adjust the presets for each one. You can make 3-band EQ adjustments and adjust presence, speed, depth, and brightness. As mentioned above, there are many other effects, which you can choose from, such as the new AIR effects seen in Figure 4.12.

If you want to keep sounds that you like, select the Preset tab as seen in Figure 4.13 and save each preset you create.

You can use the adjustor as you did in Eleven Free to create other sounds, or, as seen in Figure 4.14, change the plug-in on the track you are recording.

Figure 4.12 Configuring AIR Distortion.

Figure 4.13 Changing to Preset to select a sound.

Figure 4.14 Changing to a new plug-in while recording.

While you can plug most instruments in and run them through effects to change their character, some are specialized with presets to enhance specific qualities of their sound. For example, for guitars, you may want to emphasize mid-range equalization settings, while with the bass guitar, you will want to focus more on the lower frequencies. Normally, the Sans Amp that comes with Pro Tools 8 LE would be used for bass guitar, although you can of course run your guitar through it as well. For this section, we will talk about the bass. In Figure 4.15, you can see the Sans Amp. Here, you can adjust the preamplifier, buzz, punch, crunch, drive, as well as a low and high EQ, and an overall level setting.

If you want to hear the preconfigured settings, you can browse through presets. You can select from dozens of already configured presets (Fig. 4.16) and also adjust or make your own presets. Now you can cycle through different sounds and check out what is available.

Once you have selected the plug-in and sound you want, you can now adjust playback, so you can hear what you are playing.

You need to adjust preamp volume. You should not raise volume with a fader, as it's used to make minor adjustments, not actual initial volume settings. Figure 4.17 shows an example of how to adjust the output volume in the Edit window.

On audio tracks, level meters indicate the level of the signal being recorded or played back from the hard drive. If you see Green, you are in good standing. If you are in the Yellow, then you are roughly −6 dB below full scale, and Red indicates that the signal is getting close to clipping. As mentioned before, clipping

Figure 4.15 Using the Sans Amp from Tech21.

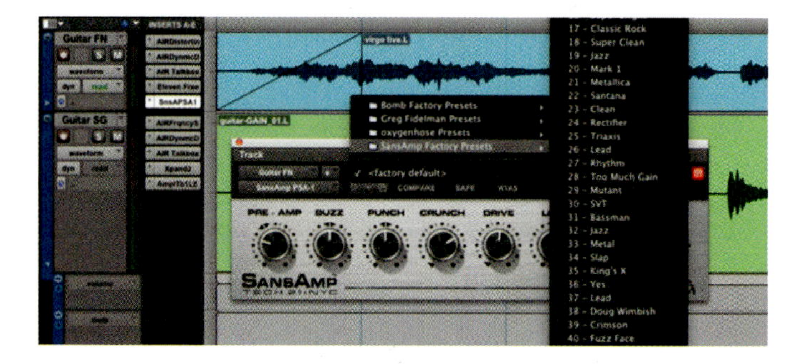

Figure 4.16 Adjusting the presets on the Sans Amp.

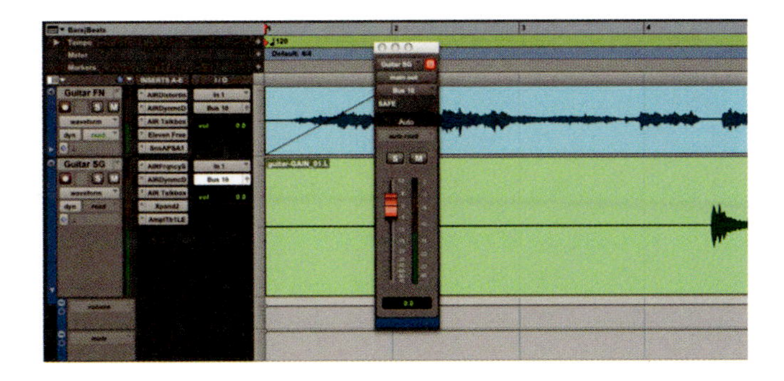

Figure 4.17 Adjusting volume in the Edit window.

will permanently ruin your recorded track by creating distortion you cannot remove. When a track is record-enabled (ready to record), the meters indicate current audio levels. On MIDI tracks, the level meter shows the MIDI level of the most recent MIDI event. You can globally set audio track level meters to indicate prefader or postfader levels by selecting or deselecting Pre-Fader Metering in the Options drop-down menu. When Pre-Fader metering is selected, this means that the level meters will show levels independent of the fader position. With postfader metering, the level meters respond to the fader position.

If clipping occurs, the light-emitting diode (LED) on the very top will turn red and stay lit. This red indicator shows that clipping has occurred not that it is still occurring. If you clip one time, you will need to adjust the volume or gain in your audio interface and re-record. You cannot remove clipping once it is recorded except sometimes with the pencil tool. You need to view the output for awhile and see how much the track is clipping. You can also view this in the Edit window, by looking directly at the waveform in the region and seeing if it's touching the top and bottom of the region, which normally indicates clipping.

Tip

To clear the LEDs from showing clipping, you can hold down the Option key on the keyboard and then click on any section of a single meter to clear all clipping information shown in the Mix or Edit windows.

Listen to playback through your monitors (speakers) or through a set of headphones. To set up a headphone mix, assign a send output to an available output on your audio interface. You then connect a headphone amplifier to the corresponding output. An Auxiliary Input send control can be used as a send master to control the overall level of your monitor submix.

Tip

You can send the mix through headphones so that the artist(s) can listen to what they just recorded. This is helpful when trying to help the artist(s) write a piece of music or get through recording a difficult arrangement. It also helps to keep the audio engineer and production staff in communication with the artist(s) just in case there are any issues, wants or needs. Savvy musicians may ask you for more effects on the overall mix. Make sure you can monitor both the playback and the live instrument.

Note

If you are not hearing audio from your Pro Tools session, make sure that no tracks are muted or soloed. If you solo an instrument that is routed to a bus and do not hear any audio, the output fader the bus is sent to may need to have its solo-safe enabled.

Tip

Remember, as we covered in Chapter 2, inserts are processed in order, so you may want to limit or gate the signal before you apply EQ to it.

4.6 Routing outputs and submixing Tracks

Once the inputs for each track have been set, you can then configure the outputs. Outputs can be thought of as "where you are sending the audio next." Most times, you will output to your Master Fader, another track, or output device. All tracks being sent to your Master Fader should be assigned "Analog 1-2" as the output. These are the standard/default outputs you should use unless you have an alternate configuration. Tracks may also be bussed and mixed through an Auxiliary Input track and then sent to the Master Fader. This is known as submixing. The bussing features of Pro Tools 8 LE give you the option to submix audio and MIDI tracks. To submix audio tracks, you can create an Auxiliary Input track and assign its output to Analog 1-2. To send the audio to the Auxiliary Input track, the outputs of the audio tracks should be set to the same bus number as the Auxiliary Input track. As you can see from working with Pro Tools, it's really just a matter of configuring your sounds, getting your tracks set up, and then using routing techniques to get different mixes, sounds, and elements into the recorded session. Submixing is just another technique you can use. It will allow you to do things such as add EQ and compression to a submix (such as a submix of several background vocal tracks) before sending the submix to the Master Fader.

You can assign plug-ins to a submix. In addition to submixing, Auxiliary Input tracks can serve as effects busses by inserting plug-ins or external effects processors. When you are submixing for reverb, delay, and similar effects processing, use a send to "bus" the audio to an Auxiliary Input track whose output goes to the same bus as the track itself. Effects setup in a send/return loop should be set to 100% wet and then mixed later with the original dry signal. The wet/dry balance can be controlled using the track's send control. A low send level sounds dry; a high send level sounds "wet" or highly effected. Before you solo any tracks in a submix, solo-safe the Auxiliary Input track so it won't mute when you solo a track – that way you can hear the effects.

You may also select Input Only Monitoring in the Track menu. To switch between Auto Input and Input Only Monitoring, either use this menu option or use the Option keyboard key with the K key. You will need to adjust this when you punch in, which will be covered later in this chapter. Input Only Monitoring means that Pro Tools 8 LE will monitor audio input when you stop playback, and when used with punching (covered later in the chapter under QuickPunch), you will be able to hear the preexisting audio until you punch.

4.7 Recording with a Click Track

A Click track is used when you want to set the tempo (usually in beats per minute – or BPM for short) for a musical selection. Many times, an artist(s) may not have perfect timing. As well, a group of artists working on the same musical selection may need help setting the tempo. Click tracks can be set to play along with the recording (in headphones) to keep the artist(s) in line with the recorded tempo. When recording with Pro Tools (or any other solution), you will be at the mercy of the artist and their performance. A Click track can be used to help keep the recording set to a specific BPM. It is critical to keep everything in tempo if possible because you will not be able to cut and paste or quantize a selection properly without having everything in line. If you intend to work with MIDI tracks in your session, or if the audio you're working with is bar and beat-oriented, you can record your tracks while listening to a click. This ensures that recorded material, both MIDI and audio, will align with the session's bar and beat boundaries.

When your track material lines up with the beats in the song, you can take advantage of some useful editing functions in Pro Tools, such as quantizing MIDI and audio regions, quantizing individual MIDI notes, and copying and pasting measures and song sections in Grid mode. Pro Tools provides a new way to create a Click track on an Auxiliary Input track by going to the Track menu and selecting Create a Click Track. Figure 4.18 shows an example of the default track created.

Tip

Beat Detective (covered in the next chapter) will help you edit your work. This is where getting a performance as close to the Click track becomes extremely important.

Figure 4.18 Viewing the new Click Track.

You can also configure the click settings within your inserts. By clicking on the Click effect, you can set the MIDI instrument that the click plays or the click accent (where the click will be heard in the count). Figure 4.19 shows the adjustment options of the Click Insert.

Figure 4.19 Adjusting the accent on the Click Insert.

Pro Tools also provides options and controls for playing a click using Audio or MIDI notes. To configure, access the options for the Audio or MIDI note by clicking the Setup drop-down menu and choose Click. You may also double-click the Click option or Countoff button in the Transport window.

You will also want to select whether the click is heard "During play and record," "Only during record," or "Only during countoff." If using a countoff, specify the number of Bars to be counted off. Countoff can help you begin the recording because the artist will have a feel for the tempo one bar (or more) before they begin. This allows for a smooth start into the musical selection. Figure 4.20 shows Click/Countoff Options, which can be invoked from the Setup menu by clicking on Click/Countoff.

Figure 4.20 Changing Click/Countoff Options.

You can also make real-time Tempo and Time changes by going to the Tempo and Time Operations windows. In Figure 4.21, you see the Tempo Operations window. Here, you can make a track selection, then choose a tempo selection change, and apply it in real time. In the Event menu, you will find the Time Operations window, which will help you adjust the timing information in the session.

When opening a new session in Pro Tools, the meter sets to 4/4 time, which is the standard default. If you need to set the default meter for a session or adjust it in any way, you can change it by double clicking the Meter button in the Transport window. The Transport window (Fig. 4.22) is where you can make more additional adjustments to your time and tempo.

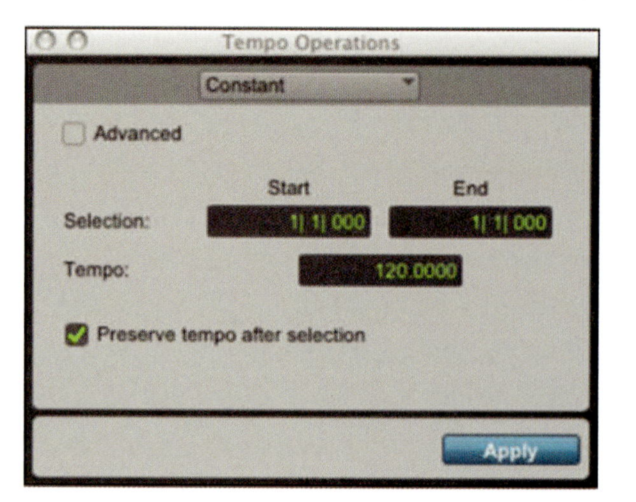

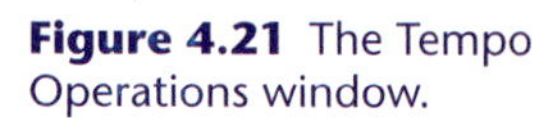
Figure 4.21 The Tempo Operations window.

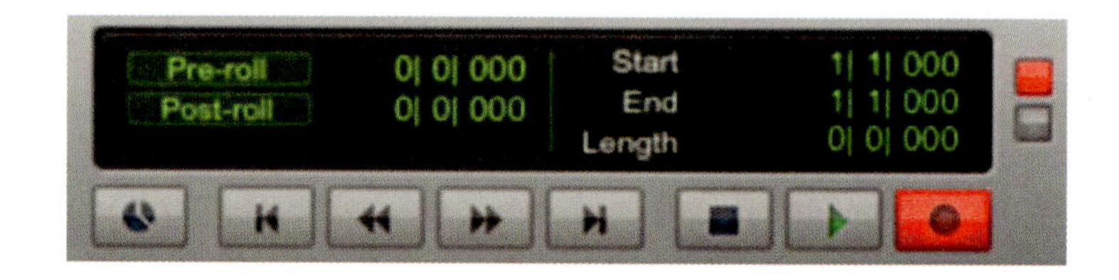

Figure 4.22 The Transport window.

Remember, you can expand the Transport area with the drop-down arrow menu and add a counter and adjusters for Pre-roll and Post-roll both of which will allow you to give a few bars before or after the section starts giving you a second to prepare of the take. You can add Pre-roll and Post-roll by selecting each (highlighting them in the toolbar) and typing in a time adjustment to the right of them. You will have to activate the MIDI Controls and Expanded Transport options (Fig. 4.23) to access these tools.

In the MIDI Controls, you can adjust the BPM of the tempo. When opening a new session in Pro Tools 8 LE, the tempo defaults to 120 BPM and a 4/4 time signature. To the change the default tempo of the session, double-click the default tempo marker located in the Tempo ruler of the Edit window. Enter the BPM value you will use for the session.

Tip

Tap Tempo can be used to set the tempo in a Pro Tools session. Click in the Tempo field, so it becomes highlighted, and tap the "T" key on your computer keyboard repeatedly at the new tempo. You can also click in the Tempo field to highlight it, and then tap in the specific tempo you want by playing it at the tempo you desire on your MIDI Controller.

Figure 4.23 Adjusting the viewable Transport tools.

You can also adjust these settings manually with the Settings adjustments seen in Figure 4.24. Here, you can adjust the Bars|Beats, adjust the tempo and meter, and work with markers.

Figure 4.24 Changing the Tempo, Meter, and Markers in the Edit window.

4.8 Creating Memory Locations

Session markers (or Memory Locations) are useful for separating and organizing data on a track in the Edit window. For example, you can note the start of a song by placing a marker. You can use the toolbar below the main timebase to add a marker. Simply clicking on the (+) plus sign to create one. Your current placement on the timeline will dictate where the new marker will be placed.

Markers are viewed in the Edit window on the timeline. When you open a New Memory Location in the Edit window, you will be able to name it, number it, and provide other details about it as well as comments. You can create up to 999 Memory Locations per session. Figure 4.25 shows the New Memory Location dialog.

You can adjust a Memory Location's placement by dragging and dropping it along the top of the timeline. Figure 4.26 shows an example of a Memory Location on the timeline. You can click it and hold it down to drag it left to right and readjust its location. You can pull it down directly to remove it completely.

Figure 4.25 New Memory Location dialog.

Figure 4.26 A Memory Location on the timeline in the Edit window.

Tip

You can also use markers to get around quickly in your session by jumping from marker to marker – say, from Verse 2 to Chorus 3. You can do this by using the period key on your keyboard's numeric keypad and the memory location number.

As we covered in chapter 2, you will need to adjust your Pro Tools Preferences in order to use this shortcut. In the Operation tab, you can set the Numeric Keypad option (or mode) to Classic or Transport. When set to Classic mode, press the number of the Memory Location you want to go to on your Keypad and then click the (.) Period. This will send you directly to that specific marker. When set to Transport, press the (.) Period, then the number of the Memory Location followed again by a (.) Period.

4.9 Setting a recording mode

You will want to set a record mode and get familiar with the functions in the Transport window before you hit the record button. You can choose between four different record modes: Non-Destructive (Fig. 4.27), Destructive (Fig. 4.28), Loop (Fig. 4.29), and QuickPunch (Fig. 4.30).

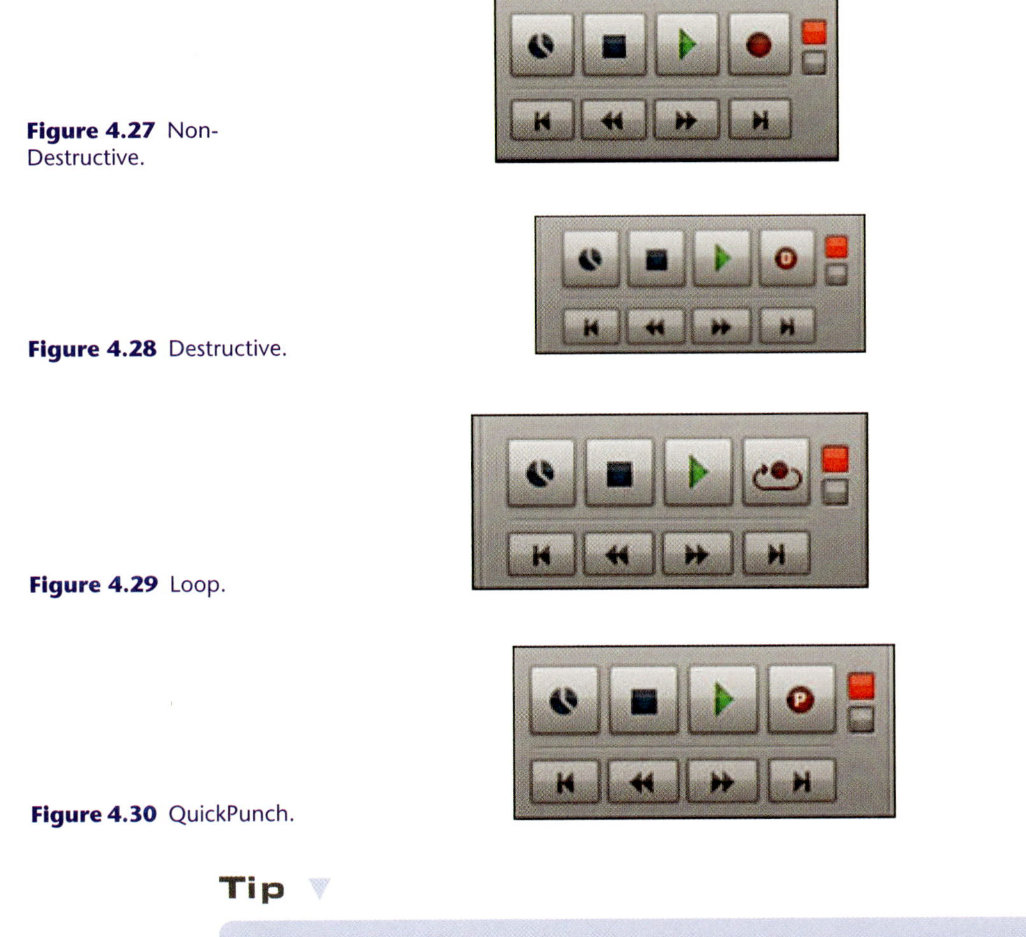

Figure 4.27 Non-Destructive.

Figure 4.28 Destructive.

Figure 4.29 Loop.

Figure 4.30 QuickPunch.

Tip

You can use the Options menu or the Transport area tool to select recording modes.

Note

When the record button is ready, it will blink on and off and be highlighted red.

Each mode offers different possibilities when recording. For example, with Non-Destructive recording, you can record to a track with preexisting audio on it, and even though you appear to be recording over it, it's still saved there as a region. You can undo the change, or as you will find while editing, move the actual region to expose the other takes. This is very helpful to anyone prone to making mistakes.

Destructive recording is exactly how it sounds in that it destroys the original region which you record over, and it cannot be restored. Destructive mode's main benefit is that it will save you some disk space.

Loop recording is useful when working with and recording loops. A loop is a section of a song that you record multiple takes of by laying down multiple takes over the same section of a track. To do this, you can enable Loop recording by selecting Loop record and then selecting a specific section of audio you want to loop on the timeline of the track. Next, arm your track, click Play to record the loop.

QuickPunch is helpful when you want to record on the same track in a specific location. You can listen to the playback while you get closer to the part you want to re-record, and once close enough, click the punch-in button and record a corrected part where you punched from.

4.10 Arm Tracks and set input levels

Next, you can use the Record Enable button to arm (set) the tracks to be able to record. You can adjust the input to a moderate level, and ensure there is enough headroom to avoid clipping the track. It's important to remember that you will want to avoid clipping at all costs; it will adversely affect the recording quality. In the Mix window, adjust the track's volume and pan faders to the proper levels. These settings are for monitoring purposes only and do not affect the recorded audio.

Lastly, verify that all connections are routed correctly and that monitor mix levels are acceptable. If using a Click track, now is the time to make sure it is enabled and that the tempo is correct. The tempo may need to be tested before recording begins.

As we just learned, select your recording mode. Pay attention to how the record button is affected in the Transport area of the Edit window. As an example, you will find that by holding down the control button on the keyboard and clicking on the record button, you will cycle through all your available recording modes.

Before recording, you should verify that your audio sources are properly connected to the audio interface and that the tracks in the Pro Tools session are properly selected, armed and routed. Other things you can do before beginning is to check to make sure there are enough resources available for the duration of your recording session. In previous chapters, we covered the requirements needed for your system and Pro Tools 8 LE in general. Your DAW will need to be inspected for proper operation and tested to ensure that there will not be any unexpected failures that may have been avoided with a simple inspection and test run of the DAW. You can use tools to check current processing power. If you check in the Window menu, you can open the System Usage window and the Disk Usage window. The System Usage window will show you current processing for PCI, CPU (RTAS), CPU (Elastic Audio), and for your Disk. This is an excellent way to get familiar with your DAW while recording and see how much processing power you are using or may need. When troubleshooting, these windows come in handy to help isolate hardware-related problems resulting from the drain of your resources.

You will want to reduce the hardware (H/W) buffer size to reduce audio latency while recording. A lower setting is seen in Figure 4.31. Remember that you will get some latency in the system if it does not use DSP cards, but nothing is noticeable unless you are running extremely low on resources or you have your buffer set incorrectly. The larger the buffer size, the larger the latency.

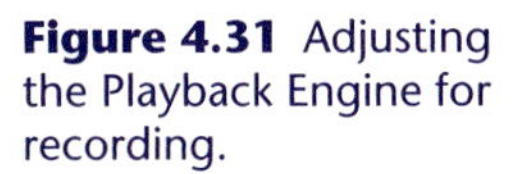

Figure 4.31 Adjusting the Playback Engine for recording.

Since Pro Tools 8 LE uses your computer's CPU to handle all audio processing tasks such as playback and recording, you will experience a minimal amount of latency or delay. One thing you should consider when using a smaller buffer size is that you will hit a limitation when adding audio tracks, the smaller the buffer size, the less tracks you can use. Problems with performance can be expected if this is not configured correctly, so if you do find an issue, you may want to readjust this buffer size as a starting point for your troubleshooting.

Tip

> Now that you have learned about the Playback Engine and how and when to use it (Chapter 2), we will use it here while working and recording. When recording and mixing, try the following preferred setting. Use a lower setting (128) when recording and a higher buffer size (1024) when mixing. You may find that a setting of 512 works great for recording about a dozen tracks with minimal plug-ins in use.

4.11 Recording audio

Before recording, you will have to consider the source. Is it a mono source or a stereo source? If it's a mono source, you will want to record to a single, Mono Audio track in Pro Tools such as the Mono Audio track we first learned about in Chapter 2. When you create a mono track and record to it, a single, mono audio file is created and written to disk. You can find the region where it appears in the playlist and in the audio Regions List.

A single, mono audio file is written to disk for each channel of a Stereo track: one for the left channel and one for the right channel; regions appear in the playlists for both channels. In addition, a stereo region appears in the audio Regions List. Figure 4.32 shows the use of a Stereo track where you can adjust your panning from left to right.

Figure 4.32 Configuring panning on a Stereo track.

In the Transport window, click Return to Zero so the start and end times are cleared. This will give you a clean starting point to begin with. It will also help to verify that you are starting at the beginning of the track you are looking to record.

Click Record in the Transport window to enter Record Ready mode. The Record button flashes and indicates that you are now ready to begin. Click on Play with the recording track armed. You will now be recording audio on your track as you play it. You may have to stop and start the recording a few times (press the keyboard's spacebar) and adjust some of your settings that we just mentioned.

To play back the audio track, click Play in the Transport window to start playback. Once you've recorded an audio track and the Transport is stopped, you can undo the previous recording you just captured. To redo (or undo) an audio recording, click Edit and then Undo Record Audio once the recording process has officially been stopped. If the recording process has started and you need to stop it, a quick way is to hold down the Command key and press the period either on the numerical number pad or on the keyboard directly to the upper right of the spacebar.

When recording a large number of tracks or channels, or playing back a large number of tracks while recording, Pro Tools may take a little longer to begin recording. To avoid this delay, put Pro Tools in Record Pause mode before beginning to record. Also, be patient, your hardware may not be able to handle the power request you are asking of it. Your DAW being configured correctly is an absolute must before you get to this point in the recording process. If you did not follow the directions in the other chapters to make sure that all your configurations are set correctly and tested, you will definitely feel the lack of preparation here when you want to begin recording. Once you have completed the recording process, you will want to save whatever was captured to your internal or external hard disks or another storage medium. It's suggested that you always save to separate 7200 RPM drive (internal or external such as a FireWire-based external hard drive) and not to the systems main hard drive, where system files are installed. Once you know how to record, you can change it by doing punch-ins, setting up loops, and doing overdubs. Recording to playlists is important to get multiple takes to use when going for a master take – or "the one."

To create a playlist you can select New... by clicking on the arrow to the right of the track name field. As seen in Figure 4.33, you can create a new track and name so that you are familiar with which take is which.

Creating a new playlist will save the older one and open a new blank one for you to use. Do not panic, your other track is safe. Simply use the down arrow to the right of the track name field and select the track you previously saved (Fig. 4.34).

This is how you can save takes and decide which to use later. Save your playlist with the proper name as seen in Figure 4.35. This reduces the need to create new tracks for additional takes, which consumes DAW power! You can do this manually – when editing, piece together the take that you ultimately want to use.

Figure 4.33 Adding a new playlist.

Figure 4.34 Selecting and replacing a previous take.

Name for duplicated playlist: Guitar SG.01

Cancel OK

Figure 4.35 Name your playlists logically.

4.12 Track Compositing

So, you love the idea of using playlists and want to combine loops in as well; this is where track compositing (a new feature in Pro Tools 8 LE) comes to the rescue. Pro Tools 8 LE offers new track compositing tools that allow you to combine the power of looping, playlists, and recording takes to secondary playlists to build a supertake. To get started, you must first set it up in your session. First, open the Pro Tools Preferences and click on the Operation tab. Here, edit the Record subsection as seen in Figure 4.36.

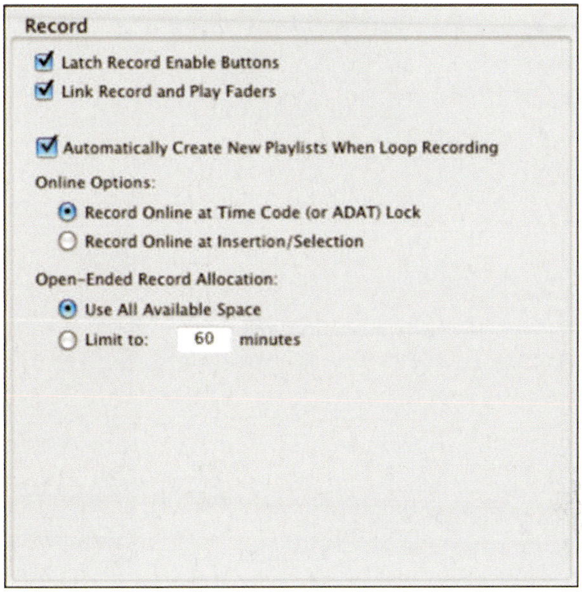

Figure 4.36 Record subsection under the Operation tab in Pro Tools Preferences.

By putting a check in the box for "Automatically Create New Playlists When Loop Recording," you will now allow new tracks to create these special playlists in Pro Tools 8 LE. That's the main change you need to make, just a few more small ones, and you can start to record takes.

Next, you need to verify that your recording mode is set to Loop Record. If you are recording a guitar solo as we are doing here and want to use the best take out of five or so takes, you record the first track and then, while loop recording a highlighted section, Pro Tools copies the takes down to the playlists on each pass. So, all that you need to do is highlight on your audio track with the Selector tool how long you want the loop to be and hit Record.

Once you click record and begin, you will not see the new playlists immediately. Each pass (or take) simply creates a new track, new region, and new waveform directly underneath the main track as seen in Figure 4.37. This will show up as soon as you stop the recording.

Now that you have five extra takes, you can listen to them and see which ones you like best. Highlight the take you want to listen to and click on the Solo button as seen in Figure 4.38. Here, you can audition and then select the best take.

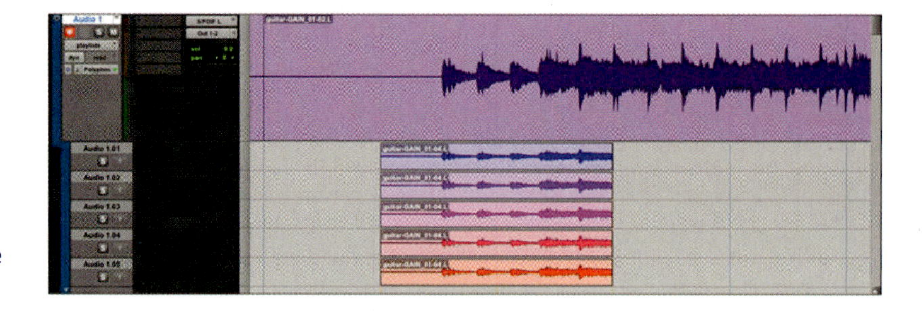

Figure 4.37 Track Compositing records multiple takes on separate tracks.

Figure 4.38 Auditioning each take with the Solo button.

Next, you want to copy the best take as a whole to the top track. To do this, you need to turn the Solo off. You will notice as seen in Figure 4.39 (on Audio 1.02) that since the region is selected in the Edit window (this is the take I want), there is an up arrow which will push that take up above all the takes in each playlist.

As you will learn in the next chapter, you can freely edit these playlists and can piece a take together using any of the tools, such as the Grabber, Zoom (or Zoomer), Scrubber, and Selector tools. Now, you have a final take that you want to keep, and it was pretty easy to get using as little manual playlist creation as possible.

Figure 4.39 Copy the best take to the main playlist.

To further edit your playlists, hold down the control key on your keyboard and click on the track title name field. This will open the menu as seen in Figure 4.40.

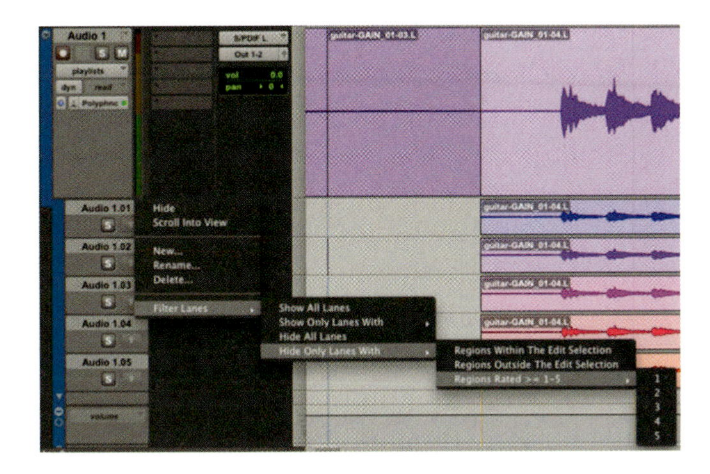

Figure 4.40 Adjusting Playlist settings.

Keep an eye on your resources while you record!

If your system is not scaled up and loaded with RAM, CPU power, and disk space, you will feel it. Your sessions may hang and freeze, and you may even lose some of your work. You can keep tabs on your system's resource usage by using the Disk Usage and System Usage windows, as seen in Figure 4.41.

The Disk Usage and System Usage windows are also customizable. As you can see, if you use the drop-down arrow on the top right-hand side of the dialog, you can adjust between a text-only version, as seen in Figure 4.42 or one with a meter.

You can also hide the activity details and just leave the meters showing if you know by memory which one is which. With Pro Tools 8 LE, you will always be concerned about saving real estate space on the desktop. These windows add up!

Figure 4.41 The Disk Usage and System Usage windows.

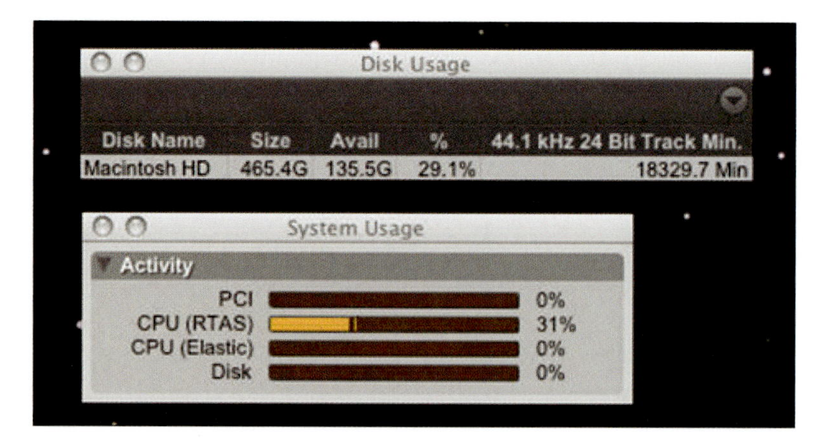

Figure 4.42 Altering the view on the Disk Usage and System Usage utilities.

4.13 Finishing the recording Session

Once the recording process has been completed and you have a recorded piece of material, you will want to save it. Good engineers will always save their work throughout the session just in case and make sure that the AutoSave options and other adjustments give you the most undo takes possible. You never know what could happen, and it's extremely important to capture all your work and have it reproducible when needed. It's also helpful to increase the history buffer so that as you work, your system contains the most information possible.

Once you have recorded your work, you can click on the File menu, and then Save to save your work. Take note of the key command for this function (Command + S), as using it after major changes or alterations or after recording is completed should become second nature. Saving your work periodically is a good practice even if AutoSave is running. To save your work from the Windows keyboard, press Control + S.

Pro Tools sessions can be saved with a different name by using the Save As command. A copy of the session can also be created using the Save a Copy In command. Save a Copy In can also be used to create Pro Tools session templates. Pro Tools sessions should be backed up regularly. We will learn how to back up sessions later on in Chapter 7.

Tip ▼

It's good to review AutoSave options in Pro Tools Preferences. You can adjust the frequency and perhaps save yourself some work.

Now that you have saved your work, you can break down. If you do not have any new takes to record, put your gear away. Every microphone sold comes with some form of case even if it's the box it's purchased in. Do not leave your microphones out and exposed to dust. Microphones are also affected by temperature and humidity. It's worth the time to break down your microphones and put them away if you think they will be ruined in any way. You want the mic to be sharp and ready to go as it's the "tip of the arrow."

Roll up any cables you were using, so you won't damage them, and wipe down your instruments, console, and whatever else you were using as you pack up after the recording session. Remember your DAW power-down procedure if you are taking a break before editing and mixing (recommended) or continue right ahead with the lessons here, and we will start the editing process! Now we get to take all the tracks we recorded (MIDI or audio) and finish tweaking them and getting them ready for the mixing process.

4.14 Summary

Once all of your tracks are recorded, it's time to edit. You spent this chapter focusing on how to get your Pro Tools session ready to record and then you captured your recorded work and saved it for editing. From here, we shift our attention to the Edit window, the main window in which you will edit your waveforms. Chapter 5 explores the use of the Edit window and what you can do with it, including learning about new features available with Pro Tools 8 LE. We will learn about regions in depth, how to use Pro Tools 8 LE editing tools to do many different things such as adding fades, using Beat Detective, removing silence from tracks, and much more.

In this chapter

5

Editing

In this chapter, we take your recorded work and show you how to manipulate tracks and regions, and enhance the recording with new tools and features found in Pro Tools 8 LE. We will learn about the Edit window, editing modes, editing tools such as the Zoomer, Smart, and Grabber, and much more. In addition, we look at how to use fades, work with playlists, and use Beat Detective.

5.1 Introduction

You explored many new areas of Pro Tools 8 LE while trying new and different plug-ins to get sounds for your recording, worked with different instruments, pickups, drum heads, amplifiers, and microphones. You worked with trying different volume settings, microphone positions, and effects. In Chapter 4, Recording, the most important thing to remember was that whatever sound you wanted to get should be captured before the editing session, not during. If you can capture the sound you want without editing it, then you will ultimately have a better recording overall and save yourself a lot of time in the process.

The editing portion of the workflow is where you clean up your work once you have completed either the composition or recording process. Whether working with audio or MIDI, you should feel comfortable with making changes to your recorded work and trying to make it better. There are obvious things you can do to your captured work to make it better, as well as things you can do to make it worse. The editing process is where you clean up your captured sounds, not alter them completely or, worse, define them. This means that ultimately you want to edit your work as little as possible and capture what you intended to hear in the recording process. Not all problems can be solved in the editing phase. It's usually the most time-consuming, especially if the editing phase is used to recapture sounds, or make new ones entirely.

As an example, we mentioned in Chapter 4, Recording, that if you are recording live instruments and did not take care of those instruments or practice with them, the performance you captured may reflect that. If your session wasn't rehearsed prior to recording, you will find the performance lackluster and will try to not only fix it here in the editing phase but even in the mixing phase, never quite fixing the original problem. A good rule of thumb is that you should always get the best sound that you can get from the source and not tamper with it too much, just accentuate it if possible. This phase should include minor detail fixing and tightening of your performance, such as quantizing your drums or auto-tuning vocal performances. A lot of work is required to clean up and enhance your recording. This is why you should stress getting great sounds and clear takes in preproduction (practicing) and then recording.

In this chapter, we cover the process of editing with Pro Tools 8 LE. In the following editing projects, we will learn the fine art of slicing and dicing audio (and MIDI) with the Edit window. As you will quickly see, there is much you can do with the Edit window to customize or alter your recorded work; it is truly amazing how deep and versatile it can be. In this chapter, we cover the fundamentals of the Edit window and how to operate within it. Moving beyond the basics, you will learn how to view track material, work with Regions, use Zoom features, use a Time Scale, and much more. As we continue to explore the Edit window, we cover making selections and playing back your audio. You will learn how to use markers, scroll selections, work with selections, and use the interface, the tools, and more. Next, you learn how to move audio within the Edit window as well as work with Edit commands such as cut, copy, clear, paste, repeat paste, and more. You will learn how to trim, move, and nudge regions (audio segments), as well as how to use the Smart tool. We have covered the Edit window in Chapters 2–4, so you should be familiar with how to open it and move within it. Let's get started and learn how to clean up our recorded work.

Note

An introduction to MIDI editing is covered in Chapter 3, Composing. Although we cover advanced MIDI editing topics here, more information about the MIDI Editor, Score Editor, and Editing MIDI and Instrument tracks can be found in Chapter 3.

5.2 The Edit window

The Edit window is where the audio or MIDI material recorded or imported within your session can be viewed and/or manipulated. Tracks are displayed horizontally in this window, and controls can be displayed or hidden as desired. With Pro Tools 8 LE, there are many new features and tools to add to the already deep toolset available in 7.4.2. You can see the new and enhanced Edit window in Figure 5.1.

Figure 5.1 Working in the Pro Tools 8 LE Edit window.

To open the Edit window, click on the Window menu and select Edit from the drop down menu. The Edit window has a main counter at the top, as well as transport controls, editing tools, zoom controls, and edit mode buttons.

As you can see, the Edit window is dense and gives you access to nearly everything you need to begin your editing work. Depending on your screen size of your monitor, you may or may not want to show all the available functionality and only take out what you need as you need it. It's recommended that you keep visible what you would use the most, which depends ultimately on how much you use the Mix window.

That being said, it's important to understand the different components of the Edit window so that as you begin to edit, you are familiar with what you have to work with and what each section is responsible for. Table 5.1 lists the main Edit window views and subsections.

Knowing what each view is and what it is responsible for will help you quickly compartmentalize each editing task (or groups of tasks) to speed up your workflow. If you want to see every view contained within the Edit window, you can go to the View menu, select Edit Window Views, and then select All. Here, you can also adjust what you do and do not see within the Edit window.

Tip

Most of the extra windows that are integrated into the Edit window are also available in their own windows, such as the MIDI Editor window and the Score Editor window. You can access both through the Window menu.

Note

It will take you some time (and a lot of effort) to navigate all of the options that you can work with within the window, so do not panic if Pro Tools is new to you.

Table 5.1 Edit window components

Transport	The top bar of the Edit window is where you access and adjust your editing toolbars, as well as adjust session parameters and control the overall session. Here, you have a main counter at the top as well as transport controls, editing tools, zoom controls, and edit mode buttons.
Project window	Directly below the Transport is the Project window. The Project window is where all audio and MIDI data are shown in track form. All tracks are located within the Project editor. The size and zoom of each track may be adjusted in the Edit window. Track automation, I/O, and plug-in data may be viewed and edited from this window as well. Here, you can work with a timeline, adjust your session's parameters further, and navigate your session.
Regions	To the right side of the Project window is the Regions view. The Regions view gives you access to and control over all audio and MIDI files in the session. Here, you can search for files by name and filter as appropriate. You can name, rename, and – if working with audio files – export files.
Tracks	The Tracks view gives you access to your tracks and control over them in a minimized view. This helps you work within sessions with a large number of tracks. There are sorting options and the ability to conceal tracks that aren't needed.
Groups	Directly below Tracks is the Groups section where you can create and use groups to further help your editing and mixing process. You can group tracks together to help you edit multiple tracks at once while working with editing tools such as the Selector. When mixing, you can group your instruments (such as drum tracks) and adjust their overall level as one unit once your individual track mixing is complete.
MIDI/Score Editor	The integration of the new MIDI/Score Editor into your Project window helps you to quickly view your MIDI tracks into an easy-to-use editing window that has enhanced tools. This also gives you quick access to the new Score Editor.

Universe View

New to Pro Tools 8 LE is Universe view, which replaces the Universe window in version 7. You can access the Universe view by going to the View menu, selecting Other Displays, and then selecting Universe. You can see Universe view in Figure 5.2.

Pro Tools 8 LE allows you to quickly move through your session with the Universe view. This view used to come as a separate window but now is integrated into the top of the Project section of the Edit window. Here, you can quickly see all tracks as they sit in the Edit window regardless of whether they are active so that you see and navigate an entire session quickly without having to use the scroll bars in the Edit window. You can use the Universe view to scroll through your session and locate sections within long sessions easily. For example, if you used markers while recording, you can now see them in Universe view, quickly scroll to them, and be anywhere within the session viewing any track within it in seconds.

Tracks appear in Universe view differently than they appear in the Project window. First, audio and MIDI tracks will be represented by horizontally colored lines. The colors will be the same as those found on your track's region color and will only change size (vertically) if they contain any automation data or if you are doing track compositing and have extra playlists under the track. Any track not containing audio or MIDI region information will not be shown in Universe view. Auxiliary Input and Master Fader tracks will not be displayed in Universe view.

If you need to hide Universe view, simply go back to the View menu, Other Displays, and select Universe again to turn it off. You can also resize the Universe view window by clicking on its border (your mouse pointer will change into a resizer). If you want to use and then hide it periodically, you can quickly resize it into view and then hide it again when not in use.

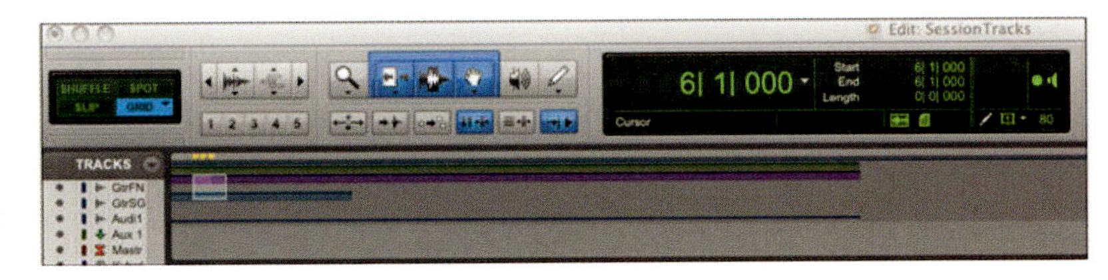

Figure 5.2 Universe view at the top of the Edit window.

Directly below Universe view is the consolidated and enhanced Main Ruler view, which you use to set your key, tempo, and chord changes as well as set memory locations (time markers). Table 5.2 lists all the major sections of the

Table 5.2 Sections within the Main Ruler	
Main Ruler (or Timebase ruler)	This is where you will find your timeline divided between Bars|Beats, Minutes:Seconds, and/or Samples. You can display all three simultaneously. "Bars" shows you the measures or bars. Each measure is divided into a specific number of beats. The number of beats per bar is the time signature. Another term for time signature is meter.
Tempo	Tempo is a term used to describe the speed or pace of the music in beats per minute (BPM). The tempo specifies whether the music is to be played quickly, slowly, or moderately. Some terms used to describe the tempo are allegro, lively, with movement, largo, and moderate.
Meter	The meter is the time signature. As mentioned, when you were setting up your rulers, you can configure this to be anything specific such as 2/3, 3/4, or 4/4 time. If using a 4/4 time signature, you will have four quarter notes for each measure, which last for four beats. You can also adjust the "click" or accented note which defines the first beat in a measure.
Key	The key is defined as the reference pitch for the song. It is commonly shown as either major or minor and can be configured for the session below the Main Ruler.
Chords	A group of notes played at the same time create a Chord. You can also configure chord changes within Pro Tools directly below the Main Ruler.
Memory Locations (or Markers)	Markers are used to show you the location of region data within the Edit window. You can use Markers to help you organize your work and locate sections within it easily, especially if you name your markers. Markers are found on the bottom of the Main Ruler and directly above your first track in the Project window.

Main Ruler and distills their use so that you can see which section handles which function.

To view (or remove from view) any section here, simply click on the drop-down stack box found in the top left-hand corner of the Main Ruler. As seen in Figure 5.3, you can add or remove any view as needed. The Main Ruler cannot be hidden.

Recorded audio or MIDI material in Pro Tools is linked to a timing reference that can be displayed as Bars|Beats, Minutes:Seconds, or Samples. This reference is known as the Main Time Scale. The Main Time Scale in the Edit window is like a ruler of elapsed time from the start of the recording. The chosen format of the Main Time Scale will affect the time format of pre-roll and

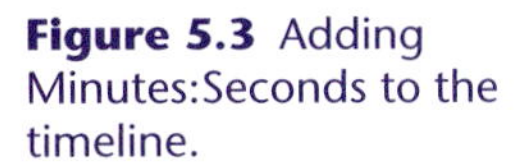

Figure 5.3 Adding Minutes:Seconds to the timeline.

post-roll amounts, start, end, and length values, and grid and nudge values as well. The Main Time Scale may be changed by clicking on the triangle to the right of the main counter in the Edit Window. It may also be changed by clicking the name of a Timebase ruler or by clicking the View menu, choosing Rulers, and then the desired Time Scale. Time Scales can be displayed as rulers at the top of the Edit window (see Fig. 5.4).

Figure 5.4 A Time Scale in the Edit window.

Proper timing is important to ensure that musical performances have a steady tempo. For this reason, Pro Tools provides many features for setting the tempo and meter of your project. As we learned in Chapter 2, Session Setup, you can configure your Main Timebase ruler as well Tempo, Meter and Key Signature rulers. These are also called Conductor tracks. Tempo and meter events affect the timing of tick-base tracks as well as provide the tempo and meter map for the Bar|Beat Time Scale. To make sure your tracks align with the bars and eats in your session, you should always record to a click. Configuring the Timebase and Conductor rulers as well as creating a Click track was covered in Chapter 2, Session Setup. Later in this chapter when we cover Beat Detective, this will become extremely important to consider.

These rulers are known as the Timebase rulers. In addition to the Timebase rulers, the Edit window contains rulers for markers, tempo, and meter as well. Rulers may be displayed or hidden by clicking on the View drop-down menu and clicking on Rulers, then selecting which rulers you wish to display. To change the order of the rulers, click on the ruler name and drag to the desired location. All the rulers may be displayed simultaneously but only one represents the Main Time Scale of your project.

Looking at Figure 5.5, you can see the enhanced view where you can now work more easily with tempo settings, changing the meter, and so on.

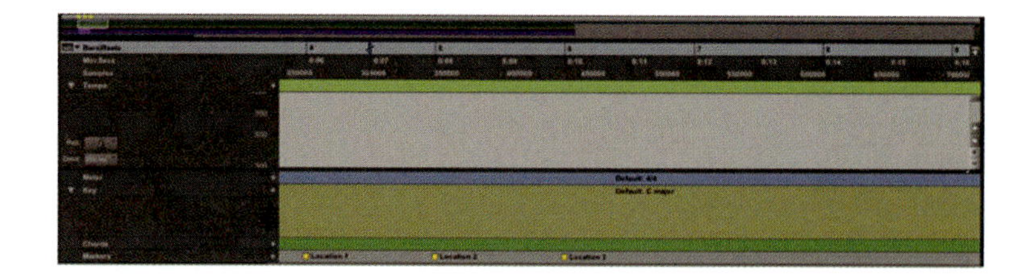

Figure 5.5 Viewing the Tempo section.

In the Main Ruler, as outlined in Table 5.2, you can set your timeline's time format to Bars|Beats, Minutes:Seconds, or Samples (you can also view all of them at once), or adjust the tempo or meter, change the key or chord as well as add markers. While editing your session, you may need access to many of these sections, especially the timeline. The Tempo Ruler can be used to insert tempo changes anywhere in the session. Each new Pro Tools session defaults to a tempo of 120 beats per minute (BPM). To change the start time of the Pro Tools session, double-click on the Song Start Marker. The Song Start Marker can be dragged to change the start point of the Pro Tools session. To insert a tempo change, click the Tempo Ruler at the desired point and then click the Add Tempo Change button at the left of the Tempo Ruler.

Note

Beats per minute (BPM) is the unit of measure of the tempo of a musical composition. 120 BPM is "faster" than 80 BPM.

Tip

Tempo changes can be moved by dragging them anywhere in the Tempo Ruler. To remove a tempo change, press Alt (Vista/XP) or Option (Leopard) and click on the tempo change. To remove multiple tempo changes, drag over them with the Selector tool (covered in the next few sections) and press the Delete key.

You can also change the Tempo Mode. In Manual Tempo Mode, Pro Tools 8 LE ignores the tempo changes in the Tempo Ruler. To enable Manual Tempo Mode, click on View, Transport, and MIDI Controls, and then click the Conductor button, so it becomes dark. Tempo may also be adjusted by using the Tempo Slider, shown in Figure 5.6.

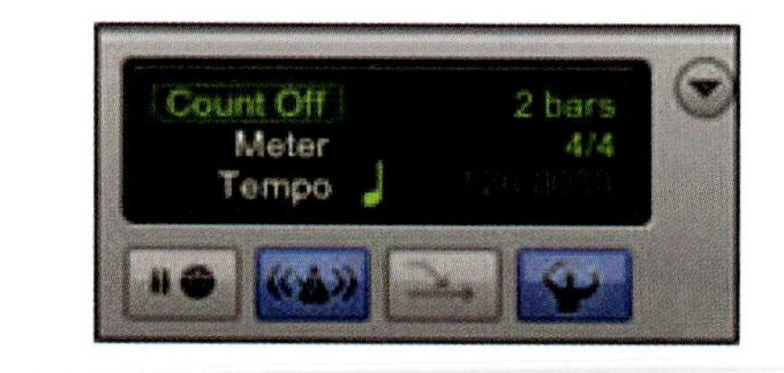

Figure 5.6 The Tempo Slider (left blue button) and Conductor button (right blue button).

The Graphic Tempo Editor allows you to visually edit the tempo changes in a session. To open the Graphic Tempo Editor, click the triangle to the left of the Tempo Ruler. In this window, you can edit tempo by dragging or trimming the tempo curve. Tempo changes can also be copied, pasted, and nudged, and new changes can be drawn in using the Pencil tool. The Graphic Tempo Editor is shown in Figure 5.7.

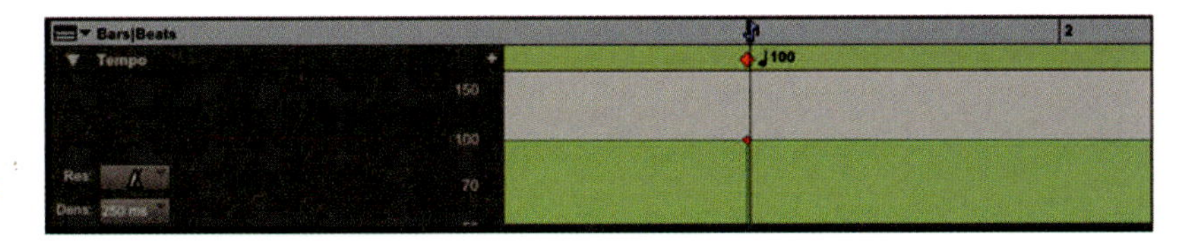

Figure 5.7 The Graphic Tempo Editor.

Tip

The Graphic Tempo Editor makes it easy to draw and edit "exact" tempo curves in Pro Tools 8 LE sessions. You can nudge and trim tempo segments to the exact tempo needed. You can use the Trim tool in the Graphic Tempo Editor to increase or decrease all the tempo changes of the session. You may want to do this if the selection is too fast or slow. The Trim tool can also be used to stretch tempo changes to cover a wider area by dragging horizontally. Either way, get used to using this tool as it help you produce whatever tempo you are looking to create and/or capture when used with Beat Detective (covered at the end of this chapter).

Note

You can open the Tempo Operations window and change the tempo on the fly by going to the Event menu, selecting Tempo Operations, and then selecting Tempo Operations window. You can also open it by holding down the Alt key (Vista/XP) or the Option key (Leopard) and clicking on the 2 found in the right-hand keypad on your keyboard. Click on Apply once you set the tempo to change it.

Not unlike the tempo track, Meter Events are displayed in a ruler at the top of the Pro Tools Edit window. The initial meter can be changed at the beginning of a session, and Meter Events can be inserted anywhere in the session. The current meter can also be seen in the Transport window. The Meter Ruler is shown in Figure 5.8.

Figure 5.8 Viewing the Meter Ruler with a 4/4 Time Signature.

To change the initial meter of the session, make sure that you are at the beginning of the session and click the Add Meter Change button to the left of the Meter Ruler. In the Meter Change window, you can specify the meter and the location of the Meter Event, as shown in Figure 5.9.

Figure 5.9 The Meter Change window.

To insert Meter Events, click the desired location in the Meter Ruler and then click on the Add Meter Change button. Enter the desired meter and location in the Meter Event window. To change a Meter Event, double-click on its icon in the Meter Ruler.

Note

You can open the Time Operations window and change the meter on the fly by going to the Event menu, selecting Time Operations, and then selecting the Time Operations window. Alternately, you can hold down the Alt key (Vista/XP) or the Option key (Leopard) and click on the 1 found in the right-hand keypad on yo ur keyboard. Click on Apply once you set the meter to change it.

Other options in the Meter Event window include Snap to Bar and Click. The Snap to Bar option will automatically move the event to the first bar of the nearest measure. The note value in the Click field represents the number of clicks that sound in each measure.

Tip

When pasting Meter Events to locations other than the beginning of the session, Partial Measures can occur. Meter Events are displayed in italics when the event falls after a partial measure.

Note

Meter is beats per measure, or also known as the time signature. To keep the meter regular, you want to duplicate the same amount of beats per bar.

A time signature, which is normally seen at the beginning of the musical selection you may be reading as 4/4, 3/4, or 9/16, sets the meter. The time signature is written as two numbers (such as 4/4), one set above the other, usually placed immediately before the first note of the selection. The top number is the number of beats in a bar. The bottom number lists the note value used to represent the beat.

If you want to change the key, you can open the Key Change dialog found below the main ruler. Here, you can adjust the key for your song to a major or minor key. Figure 5.10 shows the Key Change dialog.

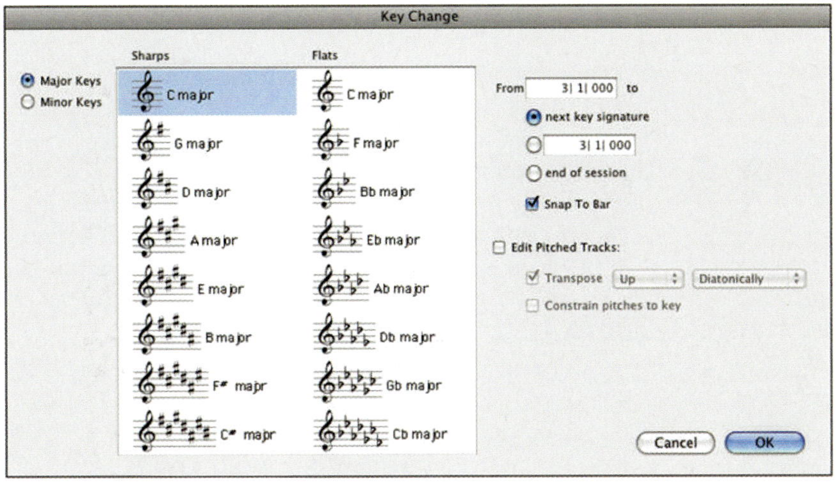

Figure 5.10 The Key Change dialog.

If you need to change a chord, you can use the Chord Change dialog, as seen in Figure 5.11.

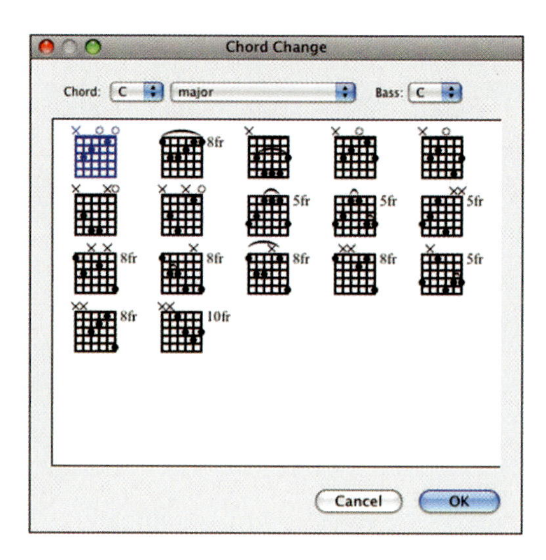

Figure 5.11 The Chord Change dialog.

5.3 Memory Locations

It might be difficult to find a certain part of your composition if the recording is long. Marker Memory Locations are useful for quickly navigating to important locations in the Pro Tools 8 LE session and are displayed in the top of the Edit window in the Markers ruler. There are three types of Memory Locations: Marker, Selection, and None. Any combination of options may be saved with a Memory Location, including zoom settings, pre-roll/post-roll, track show/hide, track heights, and group enables. In Chapter 4, Recording, we covered the creation of these markers so that you could track your work.

When creating a Memory Location, you must first select options from the Memory Location dialog. There are three Time Properties options in the Memory Locations dialog. The Marker option recalls a specific point on the timeline. The Selection option recalls a selection made in the Edit window. The None option stores only information about the General Properties options and does not recall a time or edit selection. There are five General Properties options. The Marker can be given a name and comments may be added in the field below.

Note

Before creating a Memory Location, make sure the Markers ruler is displayed and that Link Timeline and Edit Selection is enabled. When Link Timeline and Edit Selection are highlighted, edit selections made in the Edit window are mirrored in the Timebase ruler. When the icon is not highlighted, playback is not affected by the edit selection. We cover this tool in more detail in the next section.

You can create a Memory Location by selecting the desired point and clicking the Add Marker/Memory Location button at the left of the Markers ruler. You may also use the keyboard to quickly add markers by pressing the Enter key.

Note

Note: You can use the Enter key to create Memory Locations during playback. The Default to Marker and Auto-Name Memory Location options must first be selected in the Memory Locations window.

Memory Locations may be recalled and edited by using the Memory Locations window and displayed by clicking on Memory Locations in the Window drop-down menu. The Memory Locations window lists each location, all of which can be renamed. Options for General Properties may also be selected in this window. Time information can be displayed for each location, and you may add, remove, or delete locations in this window. The Memory Locations window is shown in Figure 5.12. Here, you can adjust many aspects of how your markers are displayed as well as sort, organize, and control how you view your work.

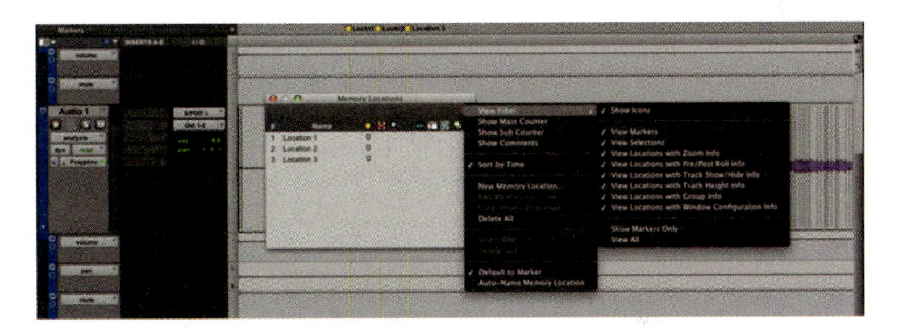

Figure 5.12 The Memory Location window.

Memory Locations may also be recalled using the numeric keypad. With the Memory Locations window open, you can quickly move throughout the session with your markers. When using the Classic setting for the Numeric keypad, recall a Memory Location by pressing the location number followed by the period. If you are using a numeric keyboard set to Transport or Shuttle, recall a location by pressing period and then the Memory Location number, followed by the period.

Tip

Memory Location markers may be moved by dragging them in the Markers Ruler at the top of the Edit window. To rename a Memory Location, double-click the marker.

As you can see, working with Memory Locations can help you quickly navigate your session, as well as further configure and organize it.

5.4 The Region List

Before you start to make edits, you need to learn about your toolset and further configure your session for editing. We just covered the basics of moving around in the Edit window as well as how to configure meter, key, and more. Now, we look at working with the Region List, how to set Editing modes, and working with tracks. When audio regions are recorded, imported, or created by editing, they are added to the Region List. MIDI regions and region groups are also displayed in this list. Files listed in the Region List can be previewed, sorted, renamed, or deleted by clicking the relevant operation in the Region List drop-down menu. Files may also be exported from the Region List menu and the full location path of the file can be displayed. Whole original regions are listed in bold. Operations from the Region List drop-down menu can be performed on multiple tracks.

Regions can also be added to existing tracks in your session by dragging the file name to the desired place in the track. Regions can only be dragged to tracks that are of the same format. Mono regions can only be dragged to mono tracks and stereo regions can only be dragged to stereo tracks. Likewise, MIDI regions can only be dragged to MIDI tracks.

Tip

To preview a region in the Region List, Alt + click (Vista/XP) or Option + click (Leopard) the desired region.

When recording audio or MIDI, you may wish to save multiple takes of material. Playlists allow you to save the recorded regions and create alternate takes without creating new tracks. Playlists also allow you to make alternate arrangements without affecting the original arrangement of the material.

New tracks contain empty playlists that audio or MIDI can be dragged or recorded to. To create a new playlist, click on the Playlist Selector of the track and then click New. By using the Playlist Selector, you may also recall, rename, duplicate, and delete playlists. Pro Tools allows you to copy and paste between playlists, which is useful for making a compilation of recorded takes. In addition, you may also assign playlists to other tracks by clicking the Other Playlists menu and then selecting the desired playlist.

Note

Pro Tools automatically names each region according to the track name. Each new region is numbered subsequently, and new regions are created every time material is edited.

Next, you should configure your editing process so that if you make any mistakes, you can quickly revert back to the last edit without issue. Sometimes, you may find yourself making many, many edits and then recalling a mistake that you may have made 20 different edits ago. In some programs, the "history buffer," as it's commonly referred to, is the repository and log of your last edits in case you need to change or revert back to a previous edit.

To undo a previous operation, click on the Edit drop-down menu and click on Undo. The keyboard shortcut for this command, Command-Z (Leopard), or Control-Z (Vista/XP) is used quite often. Pro Tools can undo up to 32 previous operations. Undo operations can be redone as well. Previous operations can be seen in the Undo History window. Operations in this window are time stamped, enabling you to revert to a previous time in the session. As we covered in Chapter 2, Session Setup, it's wise to make sure that you have configured the maximum amount of undos allowed in the Editing tab in Pro Tools Preferences, as seen in Figure 5.13. You should maximize your undo count. If you wish, you can further adjust your Regions, Tracks, Fades, and Zoom Toggle details.

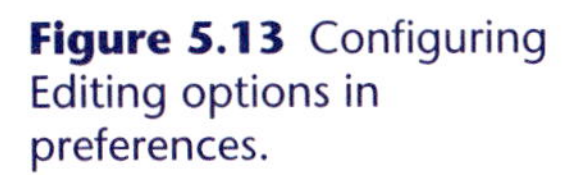

Figure 5.13 Configuring Editing options in preferences.

5.5 Edit Modes

To change the way you edit in the Edit window when working with your tracks, you need to be familiar with the main editing modes in Pro Tools 8 LE. You need to know how to operate and configure your main edit modes and the differences between them while working. Pro Tools provides four edit modes: Shuffle, Spot, Slip, and Grid. Each edit mode affects the way the audio or MIDI material of your project is arranged. The edit modes also affect the way material is copied and pasted in your session. Figure 5.14 shows the edit mode selector found in the top left-hand section of the Transport portion of the Edit window.

Figure 5.14 Viewing and setting the current edit mode.

In Shuffle mode, regions are linked continuously to other regions and may be moved and trimmed freely but cannot be separated or overlapped. In Slip mode, regions can be moved freely to any time within and between tracks without restriction. Regions may overlap or completely cover other regions in this mode. When Spot mode is enabled, you must specify the exact location of the material. Using this mode, Audio and MIDI may be placed at precise locations. When using Grid mode, regions can only be moved or edited according to the increments set in the Grid value. The grid value, which can be adjusted by Grid Value selector in the main Transport, determines the resolution or density in time of grid lines. Selecting a smaller increment of time, beat or sample allows regions to be moved to more precise locations. In addition, Grid mode may be set to either Absolute or Relative mode. When using Relative Grid mode, regions may be between beats and can be nudged according to the Grid value. Absolute Grid mode automatically moves dragged regions to the nearest value in the grid. Figure 5.15 shows that by holding down your mouse button over Grid mode, you can change it from Absolute Grid to Relative Grid.

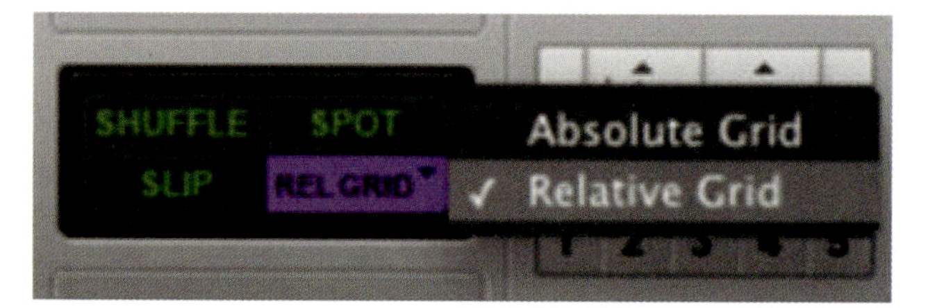

Figure 5.15 Changing the Absolute Grid to Relative Grid.

To toggle between modes, click on the appropriate button in the upper left of the Edit window. You can also use F1 (Shuffle), F2 (Slip), F3 (Spot), and F4 (Grid) to set the edit mode. Grid lines can be enabled or disabled by clicking on the name of the time format in the Timebase ruler. You can also use the Option key (Leopard) or the Alt key (Vista/XP) while selecting the 1–4 keys on the keyboard. You can toggle between Relative and Absolute Grid by selecting the 4 key twice. Knowing how to move quickly between editing modes will help you speed up the editing process, as you will find that there are specific things you can only do within specific modes. Pro Tools 8 LE has made great strides in allowing for most modes to be as flexible as possible with configurable options available for each to streamline the process.

Tip

With Pro Tools 8 LE, you can now snap to grid while in Shuffle, Slip, or Spot mode. This places the Edit cursor and makes editing sections much more flexible. To enter this new mode, click on Grid and then hold down the Shift key on the keyboard and click on either Shuffle, Slip, or Spot. You will then see both of them active for use.

You can see the Transport area along the top of the Edit window or find it in its own window called the Transport window (Fig. 5.16). You'll find a helpful tool called Big Counter in the Window menu, which is a large-scale version of the main time clock for those who have a hard time seeing or do not have a control surface handy. The playback location is displayed in counters and indicators in the Edit window Transport window and Big Counter. Now that you know the main areas of the Edit window, let's start navigating it and using the available tools to slice and dice audio or MIDI.

Figure 5.16 The Transport window.

5.6 Working with Regions

When you record or import audio or MIDI data into Pro Tools, a region is created. Audio regions are a graphic representation of audio material, shown as a waveform. Audio waveform regions display the start and endpoints of the audio, as well as the amplitude, or loudness of the recorded signal over time. See Figure 5.17, for example, which illustrates an audio waveform.

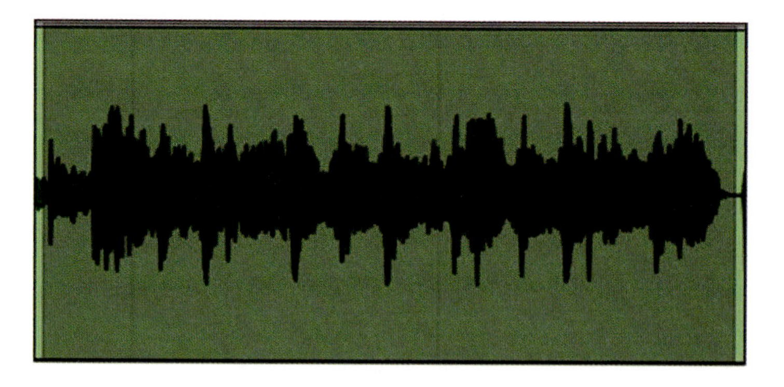

Figure 5.17 An audio waveform.

Single audio tracks may contain multiple regions, and new regions are created for each recording take. When editing audio, Pro Tools creates new regions and leaves the original file untouched. This is known as nondestructive editing. Creating tracks were covered in Chapter 2, Session Setup, as well as Chapter 3, Composing. The majority of editing that is done on audio tracks when using Pro Tools 8 LE is nondestructive. This means that if you are performing such tasks as trimming, cutting, pasting, or separating, the original source file remains intact – this is one of the biggest advantages of digital versus analog editing. Pro Tools creates a new file for the edit and the source audio files remain untouched. If you were using an analog tape recorder, you have to cut and splice the tape as you make your edits. Now you can quickly edit any data loaded within Pro Tools and not worry about losing the source of your work. Knowing how to work with regions is the key to successful editing sessions with Pro Tools.

Audio waveforms provide information about the sounds that you record. Throughout the waveform, you will see a series of peaks and valleys that represent the loudness as well as the attack and the decay of the signal. As the recorded signal gets louder, the height of the audio waveforms will increase. Waveform peaks indicate the attack of a signal and prominent peaks are known as transients. A transient is nothing more than a quick, momentary peak in audio signal level. Waveforms will appear differently for each instrument or sound that is recorded. Vocal recording may produce waveforms that have peaks and valleys that are less pronounced than instruments such as the drums. A clipped signal will extend to the top of each track and may appear to be squared off.

Once you create the tracks you need to work with, you can begin to edit them within the Edit window. The Edit window provides various features for you to display and analyze regions. The height of the tracks can be changed to show more or less detail. To change the height of a track, click on the triangle button to choose a track height from the pop-up menu, as seen in Figure 5.18.

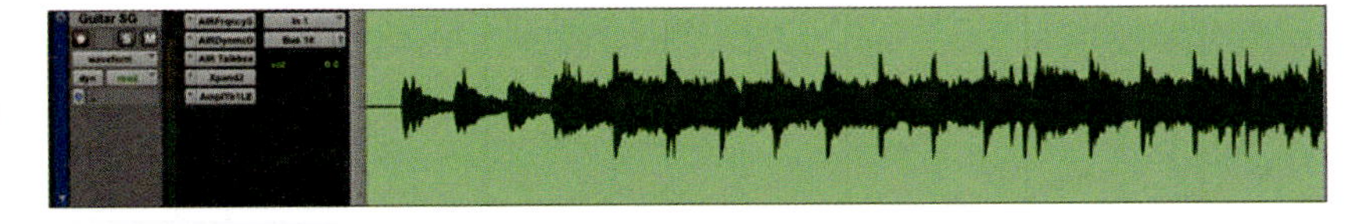

Figure 5.18 Adjusting the height of a track in the Edit window.

To change the height of all the tracks, hold the Option key (Leopard) or the Alt key (Vista/XP) and then change the track height.

In addition to audio waveforms, Pro Tools allows you to choose other types of data to display in a track by changing the track view. Audio tracks are set to waveform by default, but you may also display Volume, Pan, Mute, or Automation data. To change between Track views, click on the Track view option for that track and then choose the view that you want from the pop-up menu. MIDI tracks may be viewed as Regions, Notes, Blocks, Volume, Pan, Mute, Pitch, Velocity, Bend, Aftertouch, Program, or Sysex data. You can also expand your Track view so that you can make edits easier by either making the track itself larger for view, or using subtracks to show other information you may want to view while editing, such as panning and volume information. These settings were covered in Chapter 2, Session Setup, where you initially learned how to make and configure tracks.

Tip

In some instances, you may want to either enable or disable the display of region names and times. To display or hide specific region information, click on View, Region, and then the appropriate data you wish to display or hide.

You can see multiple examples of Regions in Figure 5.19. These examples show multiple track types with different types of information displayed.

Learning to use regions will enhance your Pro Tools experience because as you get more comfortable with what you can do (and how to do it), the doors of opportunity swing open. Using regions correctly will help you to save time and work more efficiently. When recording either MIDI or audio with Pro Tools, it's guaranteed that you will be working within regions arranging information. Understanding how regions are created, edited, and arranged is essential to taking full advantage of the editing capabilities of Pro Tools.

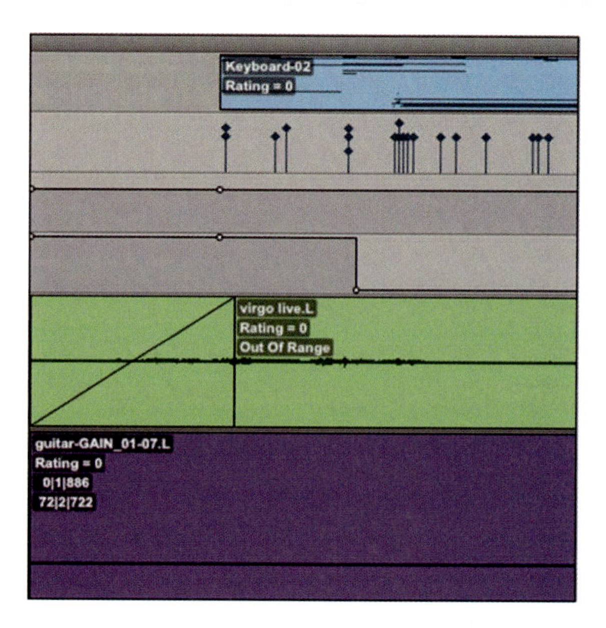

Figure 5.19 Regions as seen in the Edit window.

Tip

Before making any destructive edits to your work (which means you do not have a backup and cannot revert back to the original), you should either use playlists to save the originals or make a backup of the entire session on your hard disk. Depending on the importance of your work, you may want to save the original session, then Save As the current session with a new name if you are going to make a lot of edits.

New regions are created when recording or importing material to a track or when a region is edited. To create a new region from a selection, use the Capture command from the Region drop-down menu and enter a name for the region. Then click OK.

New regions are also created when using the Separate Region command. There are different options when using Separate Region: At Selection, On Grid, and At Transients. Separating the region At Selection creates a new region from the selected area. Separating the region On Grid creates new regions from the selection at the grid increments. Finally, separating the region At Transients creates regions starting at the transients of the selected material. Regions may also be separated and moved at the same time if you select the correct Grabber tool. The Grabber tools can select and move as well as separate and arrange regions on tracks. There are three modes for the Grabber tools, time Grabber separation Grabber and Object Grabber. Regions may also be separated and moved at the same time by using the Separation Grabber tool.

Pro Tools provides several commands for creating regions and region groups, each of them having a slightly different effect on the selection. When you create a new region or region group, it appears in the Region List and in the track's playlist.

The Region List pop-up menu provides commands to select, sort, find, rename, clear, time stamp, export, compact, and recalculate waveforms. You can change how your regions are viewed in the View menu by selecting Region and then Display on All Channels, as seen in Figure 5.20.

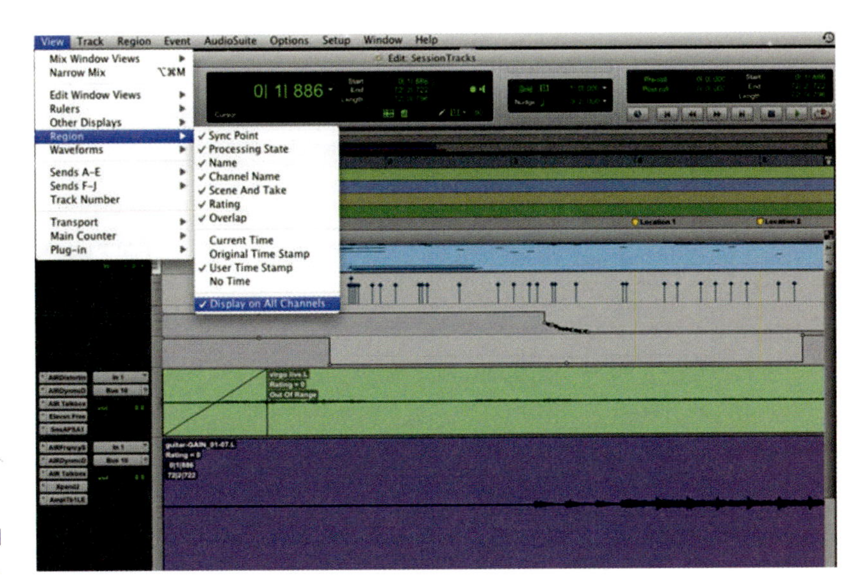

Figure 5.20 Viewing all possible Region settings.

The pop-up menu also let you set the drop order for regions dragged from the Region List and dropped in the timeline. When creating a new region from an existing region, the original region remains in the Region List. Learning to work with the Region List and creating new regions will be important as you import audio from external sources into Pro Tools 8 LE.

You can also use the Capture Region command to create a new region. This command will help to define any selection of your choosing as a new region. Executing this command will also add the new selection to the Region List. From there, you can drag the new region to any track that is already configured. To capture a new region, you can use the Selector Tool. Click and drag an existing region, and then go to the Region menu and select Capture or use the Command + R key command. Add a name for the new region and click OK. The new region appears in the Region List and the original remains that same. To work in the Region List, you can simply drag what you need from the window into the Project section of the Edit window. The Region List is seen in Figure 5.21.

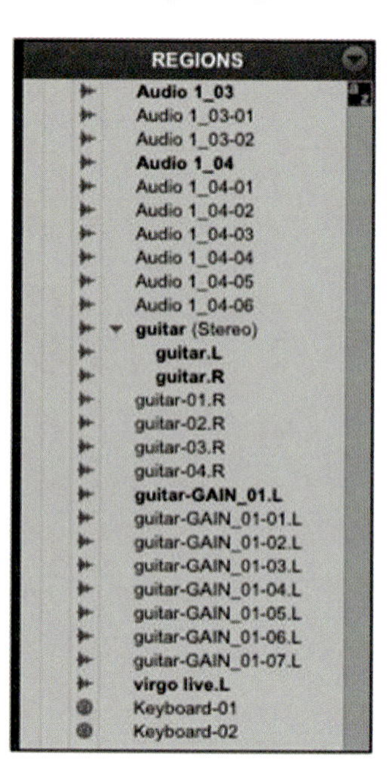

Figure 5.21 The Region List.

You can sort through and adjust the views in the Region List. To do this, click on the drop-down arrow on the top left-hand side of the Region List. This will drop down a menu, which allows you to find, sort, rename, and further edit (time stamp, etc.) your work. Figure 5.22 shows the options which you can select.

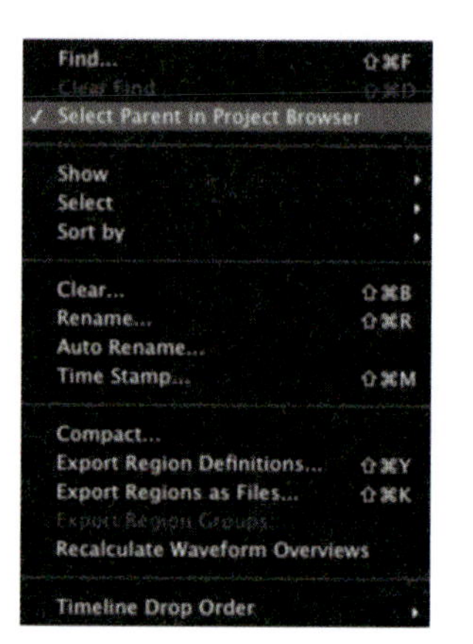

Figure 5.22 Region List options.

It's important to note that your options for working within the Region List will change as you work with either MIDI-based or audio-based regions. For example, if you choose a MIDI track in the Project window, you will not have the option to export regions as files. As we learned in Chapter 3, Composing, you can export your work as a *.mid file.

While editing, your ability to find and work with regions is critical. This is why naming them is important. To name a region, you can double-click on it and it will open the Name dialog box – here you can rename the region to whatever you wish. You can also select Rename from the Region List menu we just discussed. To find a region quickly while editing, you can select Find from the Region List menu. You can also quickly find what you need by holding down the Shift key, the Command key, and then the F key on your keyboard to open the Find Regions dialog seen in Figure 5.23.

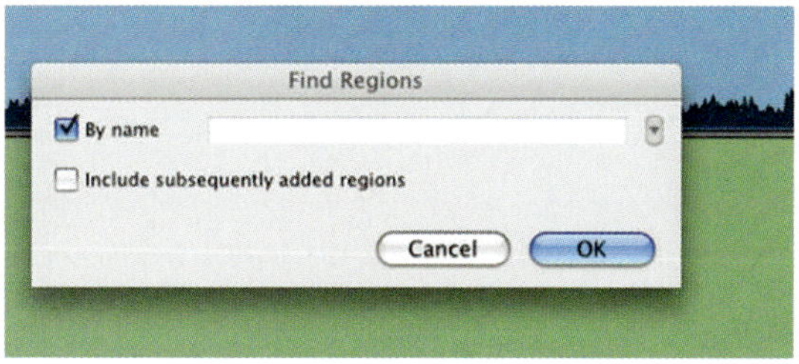

Figure 5.23 Finding a Region with the Find Regions dialog.

To start the editing process, you need to work with region data on tracks. Tracks should be loaded and ready to begin your work. To use any editing tools (which we will cover in the next section), you need to know how to work with, select, and move regions and portions of them.

Now that you understand what audio and MIDI regions are and how to work with them, let's look at working with them in the Project window. When editing, it's important to know how to select data on a track to edit it. This is where knowing your edit modes and how to read your rulers becomes very important. You can begin to prepare to cut your regions, copy sections, etc. First though, you need to know how to select a region or a portion of a region to edit it.

Apply Elastic Audio to your Regions

Working with regions has just gotten easier. With Pro Tools 8 LE, you can now enhance your editing capabilities by applying Elastic Audio to your regions as seen in Figure 5.24. You can create new tracks with Elastic Audio enabled by using the Processing tab in Pro Tools Preferences.

Now, when you want to edit (or mix), you have new ways to do both with Elastic Audio. You can pitch shift, which enables real-time changing of key. You can transpose an entire region or sections of it. There are two aspects of Elastic Audio: Elastic Pitch and Elastic Time. If you want to change either of these within your regions during editing or mixing, you should enable it on your tracks.

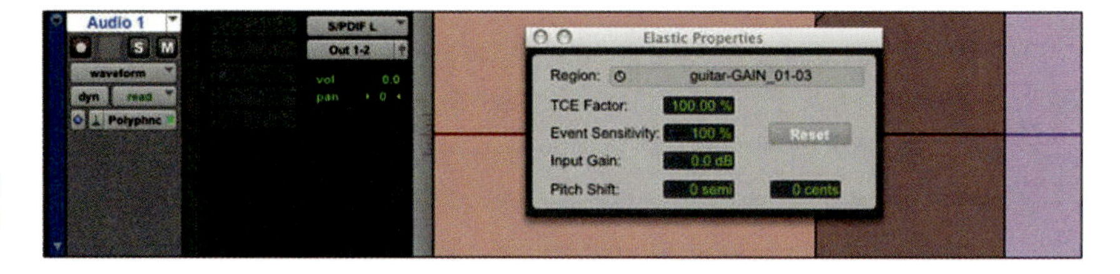

Figure 5.24 Viewing Elastic Properties on a Mono Audio Track.

5.7 Selection and zooming

When editing in Pro Tools, material will frequently have to be reviewed and played back. Material must be selected before it can be played back, moved, copied, or trimmed. There are a few ways to make selections in the Edit window using a combination of the mouse and the keyboard. When editing, you can use the mouse, the Selector tool, the Grabber tool, or the Smart tool to make selections. In this section, we explore the use of the Edit window while working with selections and playback.

There may be times while you are working when you may want to locate a point to start your playback. To do that, click on the Timebase ruler or click on the top half of a track. The Transport window may also be used to fast-forward or rewind to the desired playback point. In Nondestructive record mode, the record range can be defined by selecting a range in a ruler or in a track's playlist, or by specifying start and endpoints within the Transport window. You can also double-click the main counter and input the desired time. If the playback cursor is offscreen, the Timebase ruler shows an icon that will display the current playback time when clicked.

Material in the Edit window can be scrolled automatically during playback and recording. There are three scrolling options: None, After Playback, and Page (these can be accessed by going to the Options menu, selecting Edit Window Scrolling, and selecting the desired option). The After Playback setting displays the final playback location once playback has stopped. Page scrolling automatically scrolls the Edit window when the playback cursor reaches the right edge of the window. You can also use your mouse scroll wheel to scroll if you have one. By default, moving the scroll wheel while in the Edit window will scroll through the tracks you have created from top to bottom. To scroll from one side of the track to another (left to right), you can hold down the Option key while using the mouse scroll wheel. Once you reach the end of each track, you will then be able to shrink or expand the grid.

The most important section of the Transport section in the Edit window is the editing tools section, seen in Figure 5.25.

Caution

> When editing during playback, scrolling options should be set to None. This will keep the Edit window from following the playback cursor.

Figure 5.25 The Edit tools.

Pro Tools has several editing tools that can be found at the top of the Edit window. The editing tools include Zoomer, Trim, Selector, Grabber, Scrubber, Pencil, and Smart Tool. The Zoomer tool is used to zoom in or out of track material. The Trim tool is used to trim regions. The Selector tool is used to make selections on tracks and regions. The Grabber tool is used to select, separate, or move regions within tracks. The Scrubber tool is used to audition track material. The Smart tool is a combination tool that is used to Trim, Select, Grab, or Fade regions. The Smart tool will change depending on the location of the cursor in the Edit window. These will be covered in more depth after we finish our review of selection.

As noted earlier, there are many ways to make selections in the Pro Tools Edit window. Clicking and dragging horizontally in the Timebase ruler will allow multiple tracks to be selected. Clicking and dragging horizontally over regions using the Selector tool will also allow the selection of single or multiple regions. See Figure 5.26 for an example of a selection.

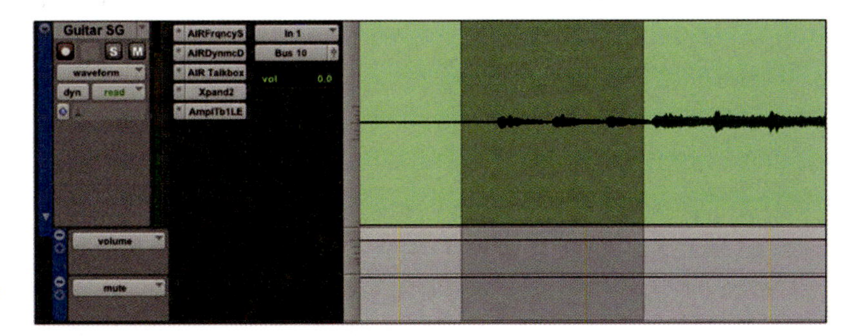

Figure 5.26 An example of a Selection.

To select an entire track, triple-click with the Selector tool in the track. To select a region, double-click on the region with the selector tool or single-click using the Grabber (the "hand") tool. Pro Tools allows selections during playback using the arrow keys. During playback, hit the down-arrow key when you want the selection to begin, and then hit the up-arrow key when you want the selection to end.

To change the length of a selection, drag the Playback Markers in the Timebase ruler or Shift-click in the track with the Selector tool. You may also use nudge to extend selections. Extend the start point of a selection by pressing Alt+Shift (Vista/XP) or Option+Shift (Leopard), and then using the plus and minus on the numeric keypad. The endpoint can also be extended by pressing Control+Shift (Vista/XP) or Command+Shift (Leopard), and then using the plus and minus on the numeric keypad. To play a selection, make sure the Link TimeLine and Edit Selection icon is highlighted, click Play or press the spacebar. See Figure 5.27 to view the Link Timeline and Selection icon.

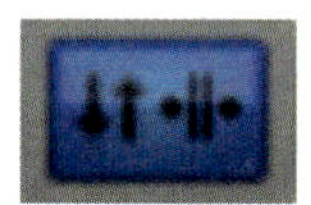

Figure 5.27 The Link Timeline and Edit Selection icon.

Note

When Link Timeline and Edit Selection is highlighted, edit selections made in the Edit window are mirrored in the Timebase ruler. Playback is not affected by the edit selection if the Link Timeline and Selection Button icon is not highlighted.

Looping may be useful when trying to rehearse a performance or when regions are being auditioned and compared many times. To loop playback, Link Timeline and Edit Selection must be enabled and a range of material must be selected. Select Loop Playback from the Options drop-down menu or Control + Click (Leopard) or Right-click (Vista/XP) the Play button in the Transport window.

To easily create a loop, make sure Tab to Transient is enabled, and click in the audio track directly before the material to be looped. Tab to the start of the section to be looped, then Shift+Tab to select the desired endpoint of the loop. See Figure 5.28 for the Tab to Transient icon.

Figure 5.28 The Tab to Transient Icon.

Once you have created your loop, you can then play it. By clicking the Play button, you will continuously loop through the selection you just made.

Note

Tab to Transient: when Tab to Transient is enabled, Pro Tools automatically moves to audio transients on pressing the Tab key. Tab to Transient is useful for placing the cursor right before a transient peak without having to zoom in on the waveform. In addition, Tab to Transient is extremely useful for making loops.

The key to a successful workflow is to be able to edit quickly. Making detail-oriented changes to waveforms as well as MIDI can take a lot of accuracy. To produce and engineer your work quickly (and efficiently), it's critical to master zooming techniques. It's recommended that you learn how to zoom in and out of sections of a track within the Edit window to help you make very detailed changes to audio waveforms and MIDI data.

The zoom features of the Pro Tools Edit window allow you to zoom in and out of your regions for either a detailed view or general overview of the material in your project. There are many ways to adjust the zoom level of the tracks in the Edit window, including the zoom buttons, the Zoomer (or Zoom) tool, or the keyboard. The Zoomer tool can be found at the top of the Edit window, as shown in Figure 5.29.

Figure 5.29 The Zoomer tool.

The Zoomer tool is used to adjust the view of the tracks in your project. Click on the Zoomer tool, and then click on the region you wish to zoom in on. To zoom out, Alt + click (Vista/XP) or Option + click (Leopard) the region. The Zoomer tool offers two modes, single zoom and normal zoom. In single-zoom mode, the tool returns to the tool chosen before zooming. In normal-zoom mode, the Zoomer tool remains selected after zooming. Figure 5.30 shows the Zoom Toggle.

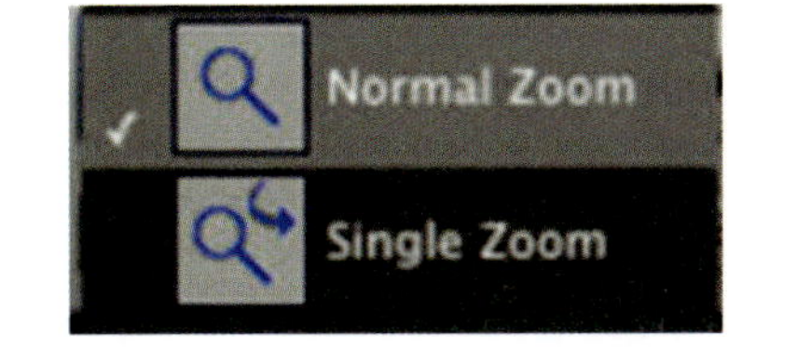

Figure 5.30 The Zoom Toggle.

To quickly change between zoom states, Pro Tools 8 LE offers a Zoom Toggle feature. This feature allows you to store the track heights, zoom settings, and grid settings of the Edit window, and to switch between them by using the Zoom Toggle button or the keyboard. See Figure 5.31 for the location of the Zoom Toggle button.

Figure 5.31 The Zoom Toggle.

To store a zoom state for use with Zoom Toggle, click the Zoom Toggle button and set the Edit window as desired. Click the Zoom Toggle button again to return to the previous zoom. To clear the stored zoom state, make sure the Zoom Toggle button is lit and then Alt + click (Vista/XP) or Option + click (Leopard) the Zoom Toggle button. If command focus is enabled for the Edit window, press the E key to toggle between zoom states.

Tip

During a busy editing session, you may find it easier to move from zooming capabilities to other tools such as the Grabber tool by using the Esc (escape) key on your keyboard. You can quickly change between the different tools in the Edit window by using the Esc (escape) key. By doing this, you can keep your right hand on the mouse and your left hand on that key to cycle through the tools as you need them while working on a specific portion of a track.

The next tool you should be familiar with is the additional zooming tools found on the Transport. Figure 5.32 shows the additional zooming tools.

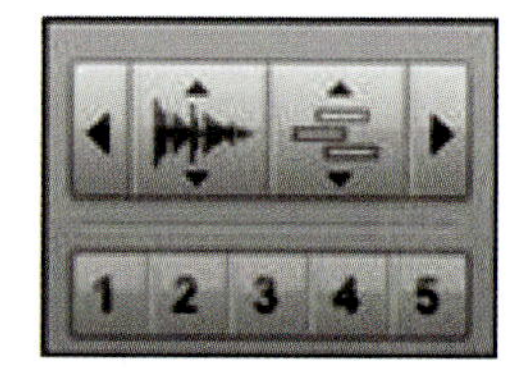

Figure 5.32 Zooming tools on the transport.

The buttons at both ends of the group allow you to zoom horizontally on regions and track data. The two center buttons control vertical zoom of waveforms in the audio regions (second from the left) and MIDI regions (third from the left). The five numbered buttons below the zoom buttons contain preset

Enhanced zooming with Pro Tools 8 LE

With Pro Tools 8 LE, you get an enhanced view when editing your regions. Because the waveforms presented on your tracks are now 16 bit as opposed to 8 bit, you get a higher resolution in the Edit window. This helps you make more detailed adjustments and helps you edit work produced from a low-signal recording. Because of its increased waveform resolution, Pro Tools 8 LE provides a cleaner palette to work on. As mentioned earlier, it's important to keep moving and stay productive while editing since there is likely much to do. Since your editing process relies on the use of zooming, it's recommended that you learn every possible way to do it, such as using keyboard shortcuts, the mouse, and a control surface. New with Pro Tools 8 LE are enhanced zooming capabilities, such as new zoom buttons located within the Project window. Directly below the Marker selection and all the way to the right-hand side directly under the Keyboard Focus selector are two handy zooming buttons that allow you to quickly zoom without having to move up to the Transport. There are also new zooming features found on the keyboard. You can quickly zoom in and out within an entire track selection by holding down your command key and then pressing on the] and [keys, respectively, to zoom in and out.

zoom levels. Preset zoom levels can be saved for quick recall by using either the numbered buttons in the Edit window or the keyboard. To save a zoom preset, set the view as desired and then click and hold one of the zoom preset buttons until it flashes.

Now that you can select data to edit and move within the track itself, you are ready to make your first edits. In the next section, we cover Pro Tools 8 LE's editing tools in more depth.

5.8 Using editing tools

Each editing tool available and introduced in the preceding sections provides a wealth of opportunities. Obviously, we have already discussed zooming and making selections. Now, let's discuss the trimming and cutting capabilities with Pro Tools. First, directly next to the Zoomer tool is the Trimmer tool (Fig. 5.33). Trimming is useful when you want to remove unwanted material from your recorded audio regions. Pro Tools provides several ways to trim audio regions through the Trim tool, the Smart tool, and the Trim Region commands in the Edit drop-down menu. The type of grid mode that is enabled affects the way regions can be trimmed. You can change the different grid views at the top of the Edit window. The Trim tool can be selected by clicking on the icon seen in Figure 5.34.

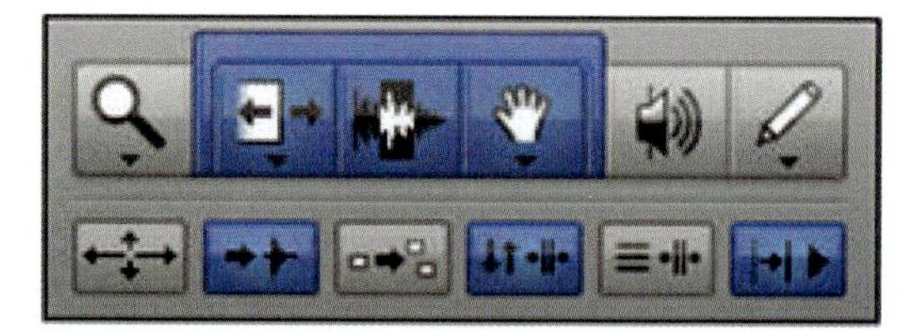

Figure 5.33 The Edit tools.

Figure 5.34 Selecting the Trimmer tool.

This tool has three separate modes: Standard trim mode, Time Compression/Expansion trim mode, also known as TC/E, and Loop trim mode. With the Trim tool in Standard trim mode, you can shorten or expand regions by clicking and dragging near the end of a region. See Figure 5.35 to view the standard Trim tool as well as the other two modes – you can open this menu by clicking and holding the mouse button on the Trimmer Tool.

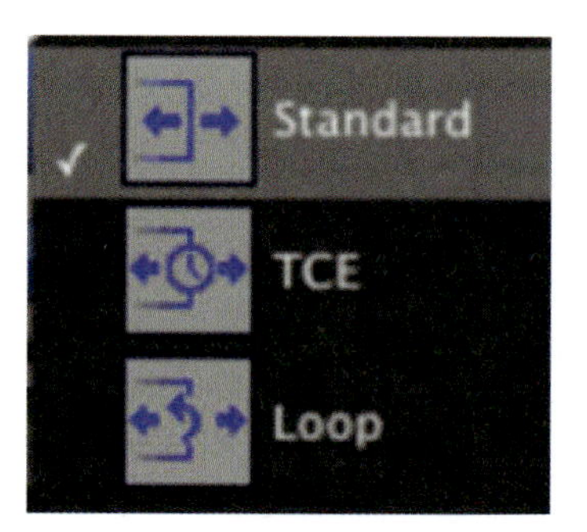

Figure 5.35 Adjusting the Trimmer tool.

With the TC/E Trim tool, you are able to adjust the region to any desired length. Settings for the TC/E Trim tool can be found in Pro Tools Preferences in the Processing tab. Here, you can select the TC/E plug-in and adjust the default settings for audio recording to any number of other presets, including film-recording presets.

Once configured, the Trimmer tool will turn into a] (trim from the right) or [(trim from the left) sign when hovered over a tracks data. The signs will change depending on which side of the center of the track you are on.

The Edit menu also contains many more commands that you can use for trimming. Here, you will find the option Trim Region to Selection. Trim Region to

Selection removes all material except for what is selected. The Trim Region to Insertion command trims material to either the start or the end of the region from the edit insertion point.

Tip

When in a hurry, you can hover your mouse over the top of your region until your mouse pointer turns into a] sign. When this occurs, you can press your mouse button and trim your selection.

Note

In the following sections of this chapter, we cover how to work with fades. You should note that if you have a fade in place, you will not be able to apply any trimming features.

Next, you should know how to operate the Grabber tool, which is located to the right of the Selector tool. You can use the Grabber tool (primarily) to move region data between tracks or from the Region List. You can also use other tools to move track data. Audio regions may be moved within and between tracks by using the Grabber tool, the Smart tool (which is a combination of the Grabber, Selector, and Trimmer tools), or by nudging. To move a region with the Grabber or Smart tool, click and then drag the region to the desired location. To move a region using nudge, select the region using the Grabber tool and press the plus (+) or minus (−) key to move the region.

The format of the Main Time Scale and the type of grid mode selected affect the way regions can be moved or nudged. Once a selection has been made, it is possible to move it by "nudging it" (moving it by a precise set amount). The results of a nudge differ depending upon whether the entire or only a portion of a region is selected and the grid type you are using. You can adjust the amount of nudging by using the Nudge selector (Fig. 5.36).

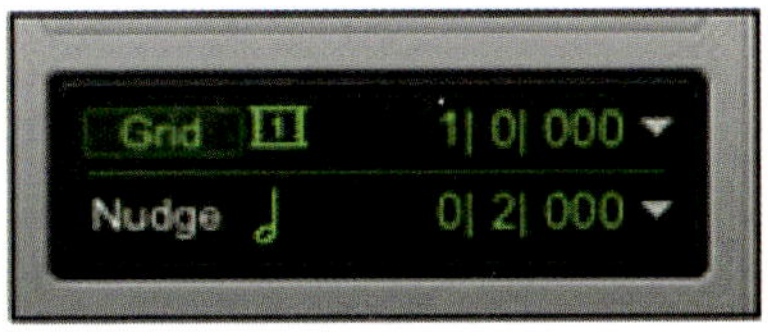

Grid 1| 0| 000
Nudge 0| 2| 000

Figure 5.36 The Nudge selector in the Edit window.

By clicking on the drop-down arrows found to the right of the nudge section counter, you can select from tempo settings. Figure 5.37 shows the adjustment of the Nudge Value from 1/4 note to 1/2 note.

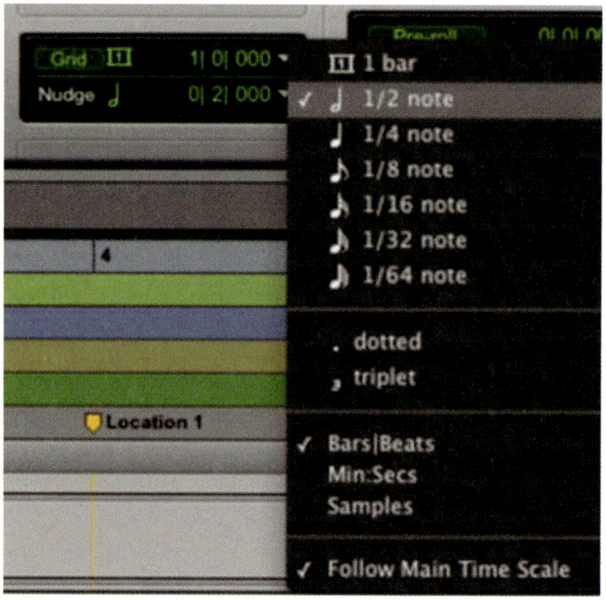

Figure 5.37 Adjusting the Nudge value to 1/2 note.

You can also adjust the amount you are able to nudge. You can change the setting at the top of the Edit window from 1 second to 500, 100, 10, and 1 millisecond. By selecting the triangle, you can change the settings of how much you can or want to nudge. When using Grid mode, a smaller time increment allows regions to be nudged or placed with more precision.

Once you are familiar with the main tools used to select, trim, and move your regions, you should also note that you can also cut them (delete them from the Project window but not from the Region List), copy them, paste them, etc. Regions may be cut, copied, pasted, and cleared very quickly by using the keyboard shortcuts. A selection must be made before the Cut, Copy, Paste, or Clear commands can be used. The Cut command removes material from the track and transfers it to the clipboard so it can be pasted to another track. The clipboard is just a storage location within your system's main memory that stores information. The Copy command places the selection on the clipboard and leaves the original selection intact. The Paste command transfers the contents of the clipboard to the edit insertion point. Finally, the Clear command removes the selection without transferring it to the clipboard. To view and use these commands, capture a new region or select a preexisting one and select one of the Cut, Copy, Paste, or Clear commands from the Edit menu.

Next, be aware that you can make your editing process easier with the Smart tool. The Smart tool automatically changes the type of tool depending on the position of the cursor over each track or region. By selecting the section directly above the Trimmer, Selector, and Grabber tools, you will "activate" all three for use at the same time. When you now move to the track and hover your mouse over it, you will see that the tool that is currently activated will depend on where your mouse is positioned. The Smart tool is used to create a configuration where when you hover over the left and right ends of the track, you will activate the Trimmer tool, when you hover over the top of the center of the track, you activate the Selector tool, and when you hover over the bottom of the track, you activate the Grabber tool. The Smart tool can be seen in Figure 5.38.

Figure 5.38 Viewing the Smart tool.

Tip

Get used to using the Smart tool. It speeds up how you can work through your editing session, as you can just move your mouse to a specific location to get the desired tool.

5.9 Creating and working with Fades

If there is one thing you will do often in Pro Tools, it's creating fades. Whether working with band recordings or producing a radio show or podcast, it's likely that your work will include at least one fade. In some projects, you will encounter, such as when creating a radio show, you may inject 200 or more fades into your project, so it's imperative that you know how to perform this function effectively. Now that you understand what a region is, it's easy to see why you would need to add fades. Consider combining different regions end-to-end to create a continuous section of audio. How could you combine the regions so that you eliminate any (or all) the noise in between each region? What if you wanted to remove any noticeable changes in sound such as clicks and pops between each region? In a radio show, you will commonly have music playing and then a DJ talking between songs and segments. What effect do you add so that you can create a nice transition from one to the other? Fading is the answer. Once the audio regions in your project have been trimmed, separated, or arranged, they may need to be faded or crossfaded to add that final finesse. Figure 5.39 shows a waveform with an applied fade. Here, it's a "fade-in."

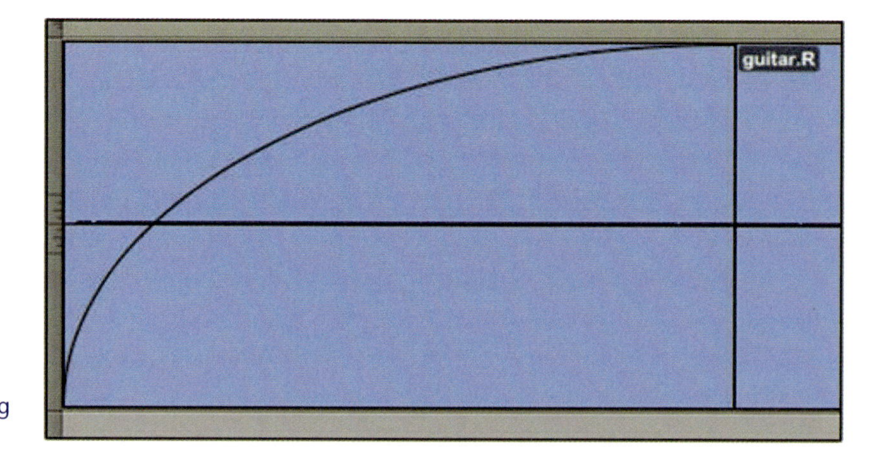

Figure 5.39 The Fading of a waveform.

When discussing fading, it helps to understand exactly what a fade is. Put simply, a fade is a gradual increase in volume when fading in and a gradual decrease in volume when fading out. You are attempting to start and end with silence. A crossfade (sometimes called a dissolve) shares the same underlying concept as a fade. In a crossfade, one track's fade-out overlaps with another track's fade-in. You can hear both tracks mixed in the middle area of the crossfade. Crossfading allows two audio regions to blend together smoothly. Fading and crossfading may also be used to prevent pops and/or clicks at the ends of regions that may be audible in the recording.

There are different ways to create fades, and the duration, shape, and placement of fades are defined in the Pro Tools Edit window. In Project 1, you will learn about the types of fades and crossfades and how to create them.

Tip

A crossfade is applied to drums when editing to help smooth the transitions between regions. Beat Detective (covered later in this chapter) will apply these when helping to smooth its work.

Fades can be customized depending on the application they are being used for. Fades can be created at the beginning or end of regions, and crossfades are created when fading between two audio regions. There are three main types of crossfades: the standard Centered Crossfade, a Pre Crossfade, and a Post Crossfade. The type of fade is determined by the way a selection is made in the Pro Tools Edit window. In a Centered Crossfade, a crossfade is created on both sides of a splice point. This is the most common type of crossfade. Both regions to be crossfaded must be adjacent to each other and must reside on the same track. A selection must be made across the splice point to

Note

Fades create fade files: when working with Pro Tools, you need to consider your hard disk space at all times. We covered the use of the System Usage and Disk Usage meters in Chapter 4, Recording. It's recommended that if running low on space, you consider using these tools to monitor your resources, including disk space. Everything you do in a nondestructive environment accumulates disk space. Fades of all types are stored on your hard disk in the Fade Files folder. The more fading you do, the more the session size will increase.

create a Centered Crossfade. An example of a Centered Crossfade is shown in Figure 5.40.

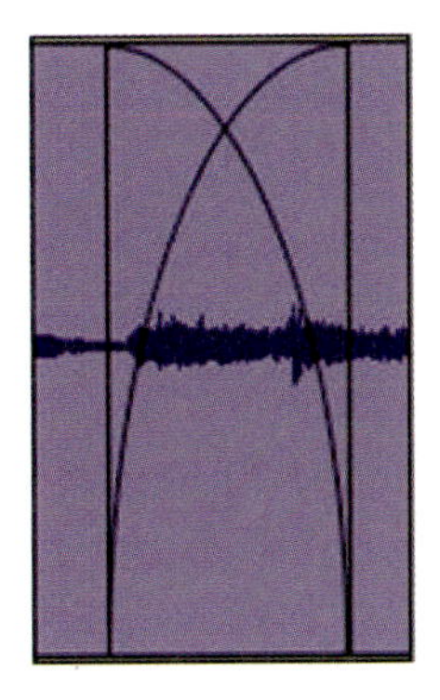

Figure 5.40 A Centered Crossfade.

A Pre Crossfade creates a crossfade before the splice point between regions. By using a Pre Crossfade, you can preserve the volume of the beginning of the second region in the fade. A selection must be made before the splice point for a Pre Crossfade to be created. An example of a Pre Crossfade is shown in Figure 5.41.

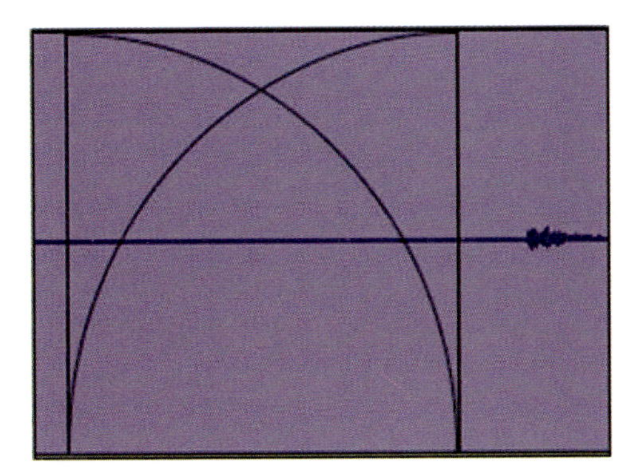

Figure 5.41 A Pre Crossfade.

Finally, a Post Crossfade is created after the splice point. Post Crossfades are useful to preserve the volume of the first region in the fade. A selection must be made after the splice point to create a Post Crossfade.

Tip

The Tab key can be used to move the cursor to the exact beginning or end of a region. The Tab to Transient button in the Transport must be disabled (not highlighted) for this to work.

Once a proper selection is made from the Edit window, a fade can be created. The Fades window is used to select, view, and preview the fade or crossfade. The Fades window can also be used to choose the type of Fade Shape used in the fade, the type of link between them, and whether or not to use Dither during the fade.

Note

Dither is a form of noise, which is added to a recording to reduce quantization distortion when converting data from a higher bit depth to a lower bit depth.

You can create a fade in a few different ways. The most common way is to open the Fades dialog as seen in Figure 5.42. To open the Fades dialog, open the Edit menu, select Fades and then Create.

Figure 5.42 The Fades dialog.

Fades can also be created in Batch Mode to create many fades at once. Create batch fades by using the Batch Fades dialog as seen in Figure 5.43.

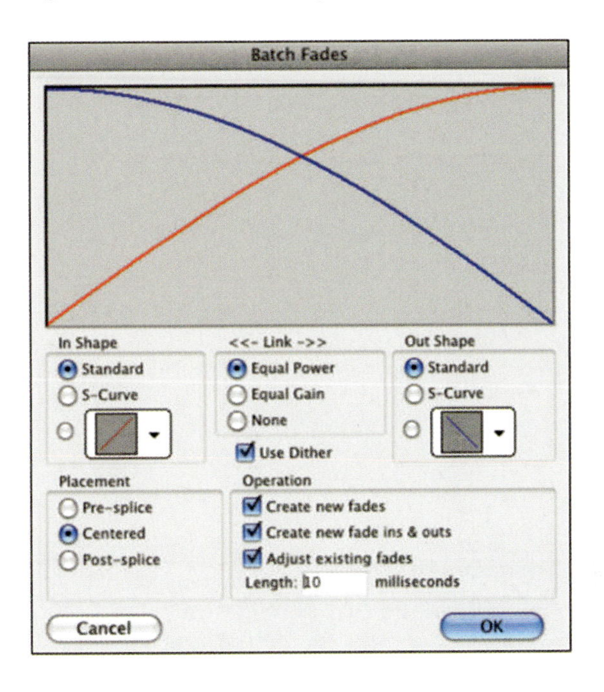

Figure 5.43 The Batch Fades dialog.

You will find a large number of extra features located within the Batch Fades dialog. If you select multiple regions at once and then open the Fades dialog, it will automatically open the Batch Fades dialog instead.

Tip

Use Command-F (Leopard) or Control-F (Vista/XP) to quickly open the Fades dialog.

There is a lot you can do in the Fades dialog. As a matter of fact, it's amazing how deep you can get into any one control on a Pro Tools 8 LE system. The Fades dialog has controls to preview the fade or crossfade with a graphical representation. To preview the fade or crossfade, click on the button in the upper left corner of the Fades dialog. To the right of the Preview button are controls to view waveforms in the Fades dialog.

The View First Track button allows you to view and preview the audio of the first pair of adjacent tracks and is enabled by clicking on the button with the waveform and the number 1. There are two of these buttons. The View Second Track button allows you to view and preview the second pair of audio tracks and is enabled by clicking on the button with the waveform and the number 2. View Both Tracks displays the waveforms of both adjacent tracks when fading more than one track.

The View Fade Curves Only button, when highlighted, shows only the fade curves without displaying the waveforms. The Fade Curves and Separate Waveforms button displays both the fade curves and separate faded waveforms. The Fade Curves and Superimposed Waveforms button displays the fade curves with a superimposed view of the faded waveforms. The Fade Curves and Summed Waveform button displays the fade curves along with a single waveform created by the crossfade.

Tip

> Fade shapes may be customized by dragging them with the mouse. Simply click on the fade line and you can move it and reshape it as needed. This can be done on the track itself or in the Fades dialog.

Options for fade shapes are available for both fades and crossfades. In addition to a Standard and an S-Curve fade shape, Pro Tools provides seven types of preset fade shapes to choose from. Figure 5.44 shows the different types of fade shapes available in the Fades dialog.

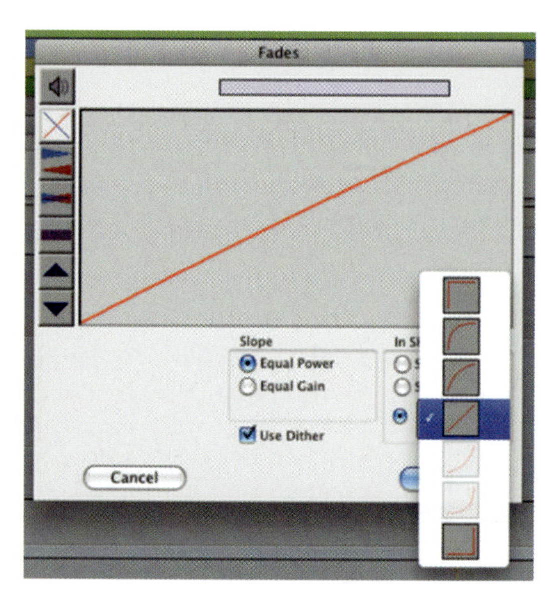

Figure 5.44 Viewing the fade shapes.

You can select these fade shapes by selecting the drop-down menu after you select the radio button. The Fades dialog also provides the following options to link fade shapes: Equal Power, Equal Gain, or None. Equal Power is recommended for a crossfade between two completely different types of material. Equal Gain is recommended for crossfades between regions of the same instrument or material. The None option disables linking between the two fade shapes in a crossfade.

To crossfade between regions, the regions must be adjacent to each other. In the Pro Tools Edit window, select the area where the regions meet. Once material has been selected, click on Edit, Fades, and then Create Fades to open the Fades dialog. Choose the type of fade shapes that you would like to use and preview the fade using the Preview button. Click OK to create a crossfade. Crossfades can also be created quickly by pressing the F key while Keyboard Command Focus is enabled. You can also use the Selector tool to select from one point of the region to the other. The selection will determine what will be faded.

Caution

Pro Tools 8 LE does not allow you to replace fades with crossfades.

To create a fade-in or fade-out, the beginning or end of a region must first be selected in the Edit window. The length of the selection determines the length of the fade. Fade-ins or fade-outs can be created by using the Fades dialog or by using the Selector tool to create an insertion point and then using the appropriate keyboard shortcut. In addition, fades can be created by clicking the Fade to Start or Fade to End command from the Edit drop-down menu. Figure 5.45 shows an example of a fade-in and fade-out on an audio waveform.

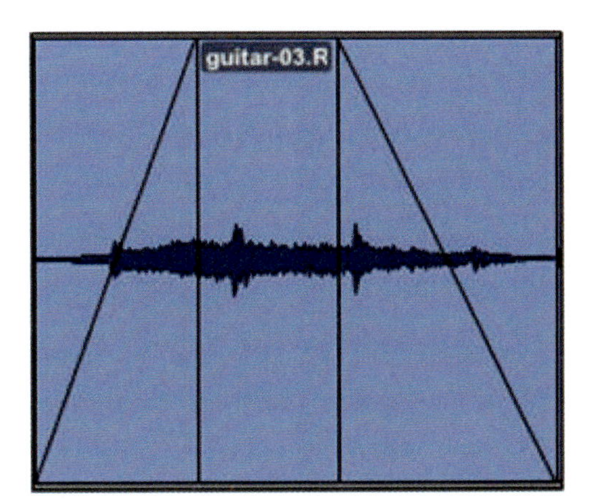

Figure 5.45 Example of a fade-in/fade-out.

Tip

When Keyboard Command Focus is enabled for the Edit window, press the D key to Fade to Start and the G key to Fade to End.

Make sure before moving on that you understand the basics of fades and crossfades, as you will use them often in your work. When working with Pro Tools 8 LE, mastering the use of fades will save you a lot of time. Make sure you practice using the keyboard to help you create your fades, it will ultimately save you a lot of time. Now that we understand the basics of fades, we will move into inserting silence into your work or removing silent areas from your work.

5.10 Using Strip Silence

In this section, you will learn about Strip Silence and how to use it. By using Strip Silence, you can quickly analyze and remove silence from selections. In some instances, you may want to edit your work so that when the musical selection starts and stops, it starts and stops from silence. Pro Tools 8 LE allows for this functionality with a tool called Strip Silence. Strip Silence is a command tool that will let you locate parts of the recorded work that are silent and strip (or remove) them from the work. This will leave your work broken into separate regions.

The silencing function acts like the Noise Gate found on most compressor/limiters sold today. One may ask "What if I do not want to remove parts that are silent?" The answer is that you can configure Strip Silence so that sensitive parts of the recording are not removed. Once understood, this highly configurable is easy to configure and useful when editing – you will be able to quickly remove silence and/or complete sections of dead space out of your work.

Tracks can be divided into regions using Strip Silence. The Strip Silence window, as shown in Figure 5.46, contains sliders for Strip Threshold, Minimum Strip Duration, Region Start Pad, and Region End Pad. In addition, there are four buttons to Strip, Extract, Rename, and Separate audio material.

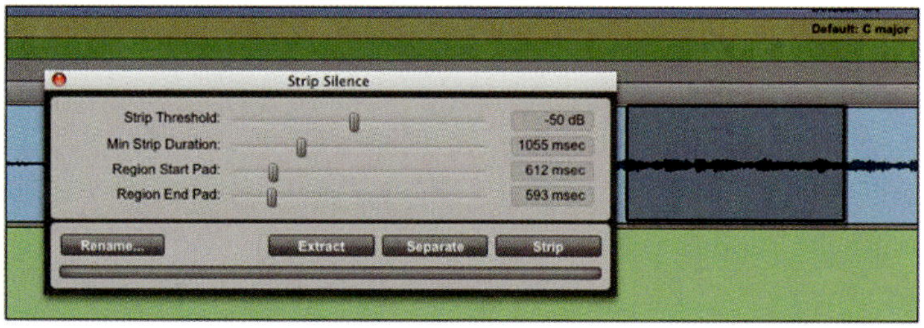

Figure 5.46 The Strip Silence window.

You can also use the new Pro Tools 8 LE "reverse" Strip Silence function, which will let you extract louder sections of your audio track. A selection must be made to use Strip Silence. If you want to choose a region, group, or regions to strip, you must use the Selector, Grabber or Smart tool and select work to be edited. Figure 5.47 shows a common warning generated when trying to strip data that is not properly selected.

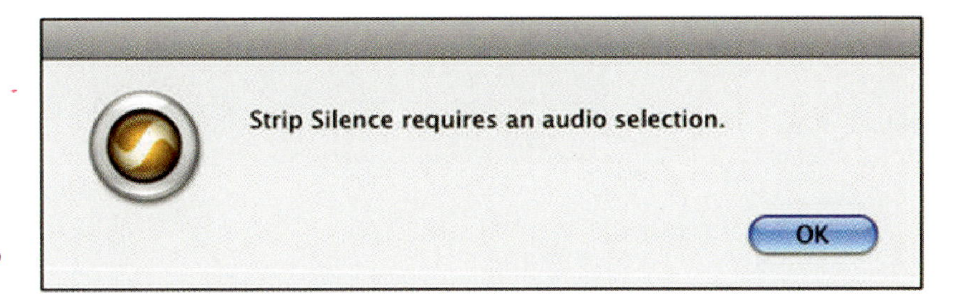

Figure 5.47 Strip Silence requires an audio selection.

Once you have selected your region, you can click on the Edit menu and select Strip Silence. You can also use the Insert Silence command in the Edit menu to create silence in the area you select with the Selector tool.

Note

With Pro Tools 8 LE, the newest feature with Strip Silence is the ability to adjust the strip threshold down to –96 dB. Prior to this latest release, you were only able to go from –48 dB to 0. Now, you can sweep from –96 dB. This is extremely helpful when working with recordings that had a low signal making the regions more difficult to edit. Now, with –96 dB, you can make your edits more precisely.

Once a selection is made in the Edit window, open the Strip Silence window and click on the Rename button to configure the naming options for the new regions to be created. When renaming for Strip Silence, you may specify the Name, Starting Number, Number of Places, and a Suffix to be added at the end of the Name. In the Rename Selected Regions dialog, the Name field specifies the root name for which new names will automatically be created. The Starting Number field specifies the first number from which new names will be numbered sequentially. The Number of Places field specifies the number of zeroes that occur before the autonumbers. Finally, the Suffix field specifies any text to be added to the end of the auto numbering.

After the Auto Naming for Strip Silence has been configured, you can adjust the Strip Threshold slider to select appropriate regions. Recorded audio material with loudness that falls below the Strip Threshold is considered silence and is thus removed. Move the slider until rectangles encompass the desired material to be kept, as shown in Figure 5.48.

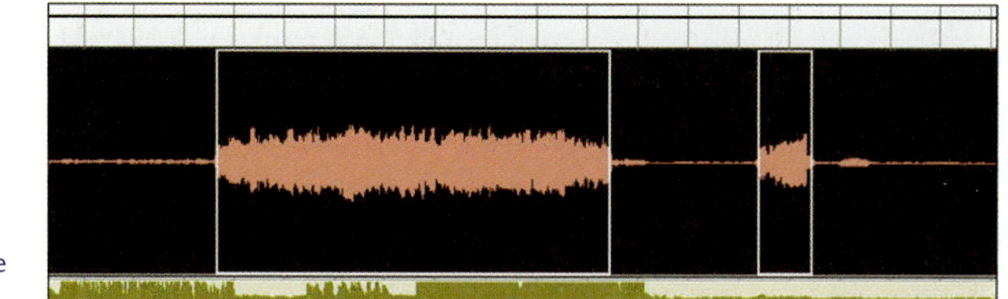

Figure 5.48 Setting the Strip Threshold.

Tip

Press Control (Vista/XP) or Command (Leopard) to adjust the sliders at a finer resolution.

Used in combination with the Strip Threshold, The Minimum Strip duration slider controls the minimum duration the audio material below the Strip Threshold must last to be considered silence and is measured in milliseconds. The Minimum Strip Duration slider may need to be adjusted to help define audio regions that are to be kept by Strip Silence. You have to be careful when using Strip Silence as it may remove work that you did not intend to be removed.

Note

The Strip Silence command is nondestructive, and audio is not removed from the original audio files. If you make a mistake, you can always go back to the original work. Although this creates more files and hence forces you to store more data, it also is a backup plan so that if you do make a mistake, you can always revert to the original session saved to disk.

In addition to removing silence, regions can be separated by adjusting the sliders to select the desired audio and clicking on the Separate button. By using the Extract button, audio material below the threshold and outside the Strip Silence rectangles is removed. Extracting is the process of removing audio and inserting silence.

The Region Start Pad and the Region End Pad sliders specify time values added to the start or end of each region created by Strip Silence. Use the slider to increase or decrease the start or end of the Strip Silence selection rectangle, as shown in Figure 5.49.

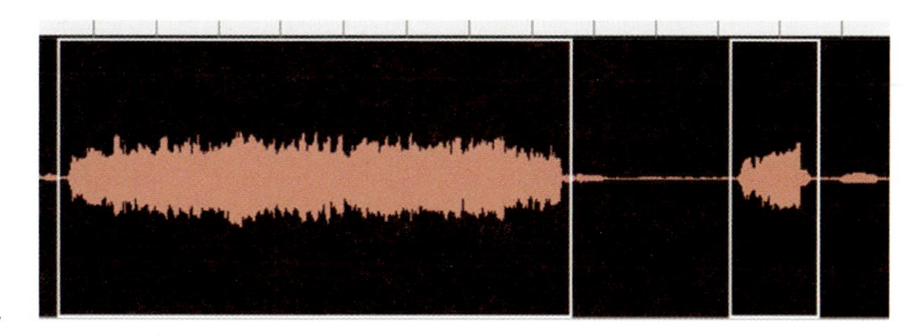

Figure 5.49 Using the Region Start and End Pad.

Once the rectangles are positioned around the desired audio material, click on Strip to remove the silence from your selection. The Separate or Extract buttons may also be used at this time if needed.

Caution

To avoid clicks, pops, or other unwanted transient noise, you may want to fade at the end of your audio regions after using Strip Silence.

Now that you have learned how to use fading to enhance your work and removed the silence from your selection, it's time to use Beat Detective to quantize your work. To quantize something means to adjust rhythm or meter inaccuracies so that any off-time or off-tempo playing is adjusted to the right tempo.

5.11 Using Beat Detective

Beat Detective is used to separate audio regions according to peak transients and quantize them according to the Edit Grid you have selected, thus improving the timing or rhythmic tightness of the recorded work. Audio may also be quantized to a different tempo (rate or speed in which something is played) and tempo may be extracted from audio by using Beat Detective. In addition, the Edit Smoothing feature of Beat Detective can be used to fade the ends of multiple regions. In this section, you will learn how to edit and quantize audio material using Beat Detective.

The way Beat Detective works is by analyzing an audio or MIDI track's rhythm pattern based on the Bars|Beats settings you have configured. Once you open it up for use, you will find many tools within Beat Detective to help you further edit your work based on beat triggers. A beat trigger is what Beat Detective processes from peak transients in your recorded work. Once these transients are located and marked, you can then further quantize, edit, smooth, and process your work.

Caution

Beat Detective cannot be used across multiple tracks in Pro Tools 8 LE. Operations of the Beat Detective window are only active in the topmost track of the Edit window. To use Beat Detective across multiple tracks in Pro Tools 8 LE, the Complete Production Toolkit (or Music Production Toolkit 2) must be installed. Installing this package (or upgrading to Pro Tools 8|HD) gives you more flexibility with Pro Tools Beat Detective.

Note

Beat Detective requires a large amount of RAM when performing its operations, especially when working with multiple tracks. As noted in the first couple of chapters in this text, it's imperative that you properly plan and configure your DAW, as not having enough system resources will absolutely cause your system to crash, be nonresponsive, or just generally move slowly, all things you do not want to happen when recording an artist. A recommended Digidesign tweak to get more power out of your DAW while using Beat Detective is to set your Undo limit to a smaller number. This will free up a large amount of memory and resources for systems that are not scaled up and powerful enough for sustain Beat Detective's load. However, you can load your machine with RAM, optimize it, and check out your System Usage meters before changing your Undo settings.

To open the Beat Detective window, choose Beat Detective from the Event drop-down menu. There are three sections in the Beat Detective window: Operation, Selection, and Detection. There are five Beat Detective modes: Bar|Beat Marker Generation, Groove Template Extraction, Region Separation, Region Conform, and Edit Smoothing. The Beat Detective window also contains information about the selected material, and detection options can be adjusted. The Beat Detective window is shown in Figure 5.50.

You can select between the Operation (MIDI or Audio), which will then give you different subsections to work with in the Selection and Detection sections of the window. Bar|Beat Marker Generation is used to translate detected trigger points to the Tempo Ruler that you selected earlier in this chapter.

Figure 5.50 The Beat Detective window.

Note

When working with MIDI and Beat Detective, note that MIDI only contains two modes: Bar|Beat Generation and Groove Template Extraction.

The Selection section of the Beat Detective window contains information about the selected material such as the Start Bar|Beat, End Bar|Beat, and Time Signature. The accuracy at which Beat Detective analyzes swing notes can be adjusted by changing the beat value in the Contains option. Click on the Capture Selection button to select material using Beat Detective.

The Detection section contains options for the detection of transients in selected audio material. The Analyze button contained in this section is used to analyze the selected audio material and generate beat triggers at each of the transients. The Sensitivity and Resolution of detection when generating beat triggers can be set in this section. In addition, detection can be set to High Emphasis to work with high-frequency instruments and Low Emphasis to work with low-frequency instruments.

The Operation Window can be further broken down into five sections for audio and two for MIDI. Bar|Beat Marker Generation mode generates Bar|Beat Markers according to the location of transients in the selected audio material. The Groove Template Extraction mode extracts the rhythmic and dynamic information from the audio material and saves it to the Groove clipboard or as a DigiGroove Template. The Region Separation mode creates new regions based on the transients detected in the audio material. Region Conform mode conforms all selected regions to the Edit Grid. In addition, audio can be quantized to the grid using either Standard or Groove quantization.

If you want to work with Groove Templates, you can select the Groove Template Extraction radio button in Beat Detective as seen in Figure 5.51.

Using this operation, you can extract your work as a DigiGroove Template. These templates can be extracted (exported) to the system's clipboard (Groove clipboard) or as a template file on your computer. These templates are helpful if you want to transfer the "feel" of the groove from one song to another.

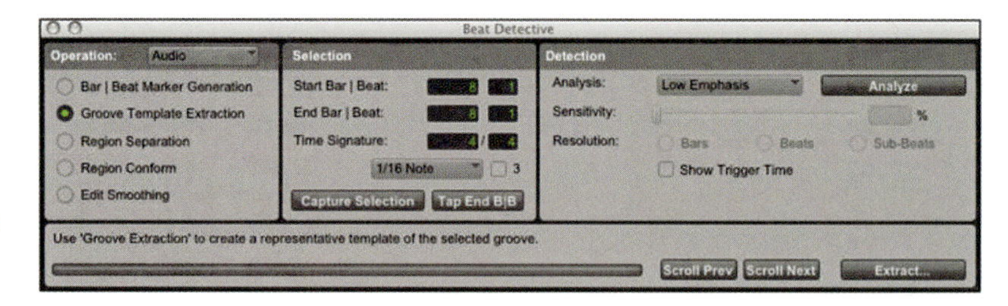

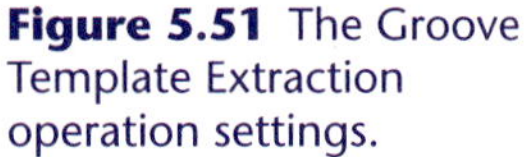

Figure 5.51 The Groove Template Extraction operation settings.

Next, you want to cut up your audio waveforms. The options in the Operation section are Region Separation, Region Conform, and Edit Smoothing. Each will be one step in a three-step process to finishing your edits. As an example, let's say you are trying to line up a kick drum track exactly to your Timebase and grid. You want to first "cut" the tracks up before and after each "hit" of the drum. The next step, "conform," is used to line up each "hit" to the sessions current Grid or to an existing groove template. Any audio region may be conformed to the Grid, but for best results, the region's start must begin at the start point of the audio material. When in Region Conform mode, regions can be aligned to the current Grid of the session using Standard Conform. As you can see, all that is left is to put the audio back together as if it was never affected. Unfortunately, the only way to do this "smoothly" is to use smoothing. This (as we mentioned earlier when discussing fades) simply puts a crossfade on all your separated regions to "blend" them together.

So, in a perfect world, if this was done step by step, you would think that your work would sound perfect? Well, not quite. Beat Detective is not the cure-all for your timing related issues. It will polish a great performance and make it sound amazing, but it will not help you with a track that is completely offtime. Beat Detective will fix it, but it will have issues where you will hear a hit of the drum that fades incorrectly or is still offtime but made worse by Beat Detective. You can, however, use the Zoomer tool to zip in and move the fades around and use the Selector tool to make small adjustments to attempt to fix it. No matter how hard you try, there will be times where you may not be able to fix a bad performance with Beat Detective. You may want to retake or rerecord the work instead, as Beat Detective can be a time-consuming process.

Note

This is why we stressed performance, practice, and using a Click track in Chapter 4, Recording. It's imperative to be as close to the beat as possible so that when you edit your work, it's not impossible to do without having to further edit the work beyond what Beat Detective can do.

Once you have selected your audio or MIDI, you are ready to edit it. By clicking on the Analyze button in the Detection section, you will start the Beat Detective analysis process. You must have your audio selected (with the audio Beat Detective also selected) or MIDI if working with MIDI tracks. This will process your work and create the associated beat triggers. Here, you can adjust the sensitivity, adjust the resolution, or configure to show the Trigger Time. Once your original selection is analyzed, you can then perform the groove template extraction. Once the Beat Detective mode has been selected, material must be captured and beat triggers must be generated. When defining selections with Beat Detective, make sure the selection starts on the attack of the first beat to ensure the best results. In addition, the selection should not contain any meter or tempo changes. Once a selection has been made in the Pro Tools Edit window, click on Capture Selection. The Start and End Points and the Meter of the selection will be displayed in the Selection section of the Beat Detective window. The Capture Selection button must be used every time a new selection is to be used with Beat Detective.

Once you have captured the selection, Beat Detective can then calculate the tempo of a selection. To calculate the tempo of a section, open the Beat Detective window and select Bar|Beat Marker Generation mode. Next, make a selection in the Edit window and click the Capture Selection button to define the Beat Detective selection.

After a selection has been made and captured, beat triggers can be generated based on transients detected in the audio material. To generate beat triggers, set the detection options in the Beat Detective window and click on Analyze. There are three types of beat triggers, and the thickness depends on whether the detected transient falls on a bar line, a beat, or a sub-beat. Beat triggers may be inserted, deleted, moved, or promoted. Some examples of beat triggers are shown in Figure 5.52.

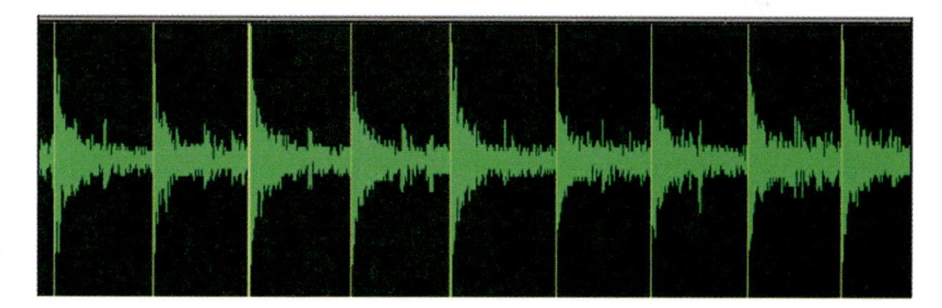

Figure 5.52 Examples of beat triggers.

Detection options may need to be adjusted and readjusted to display beat triggers in the correct locations. Beat triggers may also be edited. To insert a beat trigger, enable the Grabber tool and click on the selection in the Edit window. To delete a beat trigger using the Grabber tool, click the trigger while pressing the Alt (Vista/XP) or Option (Leopard) key. To move a beat trigger, locate and drag the trigger using the Grabber tool. To promote a beat trigger, press Control (Vista/XP) or Command (Leopard) and click the desired trigger. Promoted beat triggers will only disappear if the detection Sensitivity is set to zero.

Once beat triggers have been created, Bar|Beat Markers can be created. Click the Generate button in the Beat Detective window to generate Bar|Beat Markers. The Markers appear in the Tempo Ruler and can be used as a tempo map to which audio and MIDI regions can be conformed. Also, note that when the resolution of Beat Detective is set to detect Sub-Beats, the groove may be extracted to the Groove clipboard. In Groove Template Extraction mode, click on Extract to copy information about the tempo and dynamics and save it as a DigiGroove template.

For best results when using Beat Detective, check the location and thickness of the beat triggers to make sure they align properly with the audio material. This will prevent Beat Detective from separating regions in the wrong location. False triggers should be removed to prevent subsequent triggers from appearing in the wrong locations. Long selections may need to be broken down so that they are easier to work with.

Next, you will want to separate your work so that it can be moved around and tightened up. How is this done? Click on the Region Separation radio button as seen in Figure 5.53.

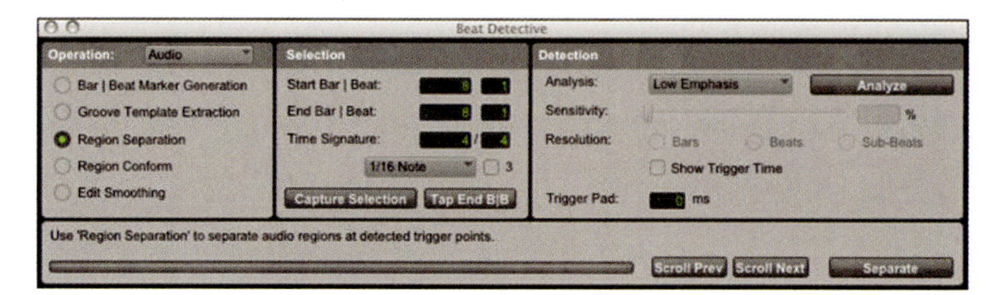

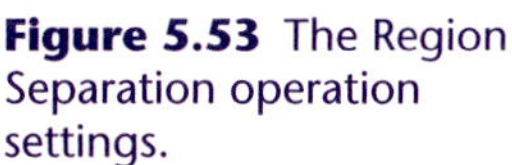
Figure 5.53 The Region Separation operation settings.

Once beat triggers are created, Beat Detective may be used to separate the selection into regions. New regions can then be quantized or conformed to the sessions Edit Grid or an existing groove template. To separate regions, Beat Detective must be in Region Separation mode. The ends of new regions can be padded similar to Strip Silence by adding a value in milliseconds to the Trigger Pad Field of the Beat Detective window. Click on Separate and regions will be separated according to the location of the beat triggers. Regions are conformed according to the location of the beat trigger, even though the region's start point may be padded.

Tip

To separate multiple tracks, extend the selection to lower tracks in the Edit window.

Conforming audio regions with Beat Detective is quick and easy. All you need to do is select Region Conform (as seen in Fig. 5.54) and then regions may be conformed to the session's current Grid or to an existing groove template.

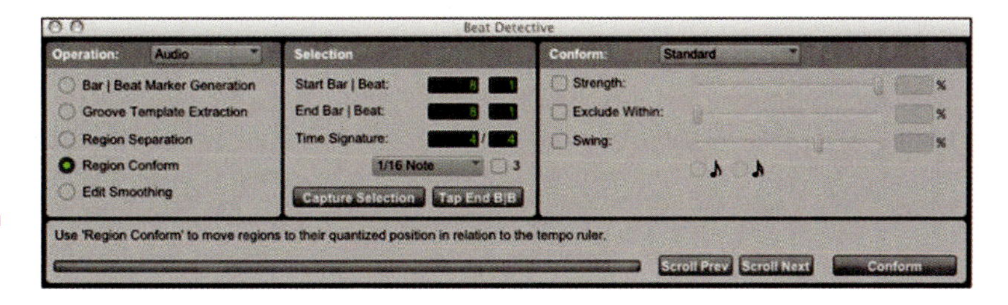

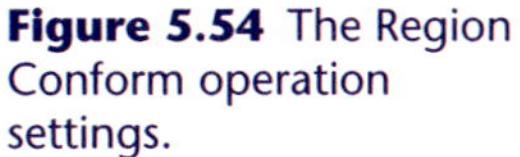
Figure 5.54 The Region Conform operation settings.

Any audio region may be conformed to the grid, but for best results, the regions must begin at the start point of the audio material. When in Region Conform mode, regions can be aligned to the current Grid of the session using Standard Conform. To conform regions, use Beat Detective to capture a selection, and then use the Strength, Exclude Within, and Swing sliders to adjust how the material is conformed to the grid. The Strength slider affects how strongly the regions are aligned to the grid. A lower Strength value will preserve the original feel of the regions and higher values will align regions tighter to the grid. The Exclude Within slider affects which regions are conformed. Lower Exclude Within values do not allow regions close to the grid to be conformed, and higher values will allow all regions to be conformed.

Tip

To conform multiple tracks, extend the selection to lower tracks in the Edit window.

The Swing slider adjusts the amount of swing feel to be added when conforming regions to the grid. The Standard Region Conform options are shown in Figure 5.55.

Regions may also be conformed to an existing groove template using Groove Conform. To conform regions to a groove template, set Beat Detective to Region Conform mode and select Groove to view the Groove Conform options, as shown in Figure 5.56.

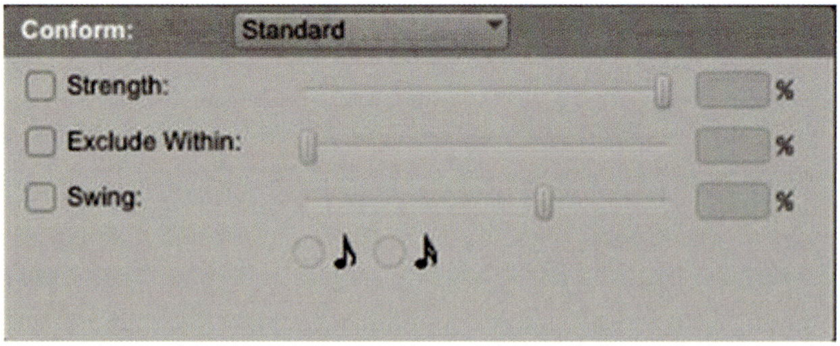

Figure 5.55 The Standard Region Conform options of the Beat Detective window.

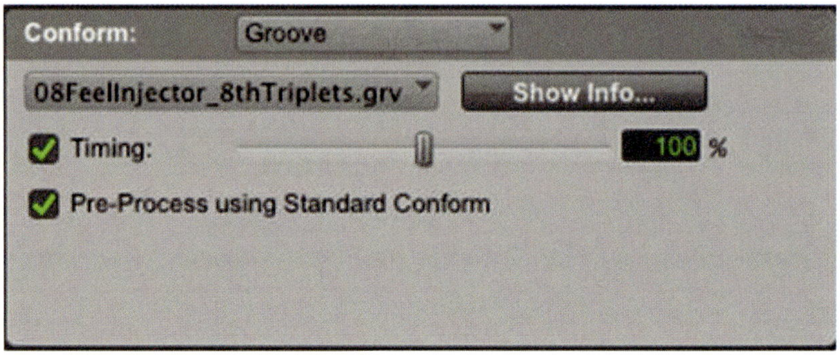

Figure 5.56 The Groove Region Conform options of the Beat Detective window.

Regions must be captured before conforming. Select the groove template from the drop-down menu and adjust the Timing slider to affect how strongly the regions conform to the groove template. The Pre-Process using Standard Conform option can be checked to ensure that the performance of the material is accurately mapped to the appropriate beats before the template is applied. Click on the Conform button to conform the regions in the selection. Once the regions have been conformed, playback the session to make sure all regions have been conformed correctly.

Caution

After conforming regions using Beat Detective, some regions may need to be adjusted manually. This is where zooming becomes very helpful.

Now you can apply fades by selecting the Edit Smoothing radio button as seen in Figure 5.57. By selecting Edit Smoothing, you can now consolidate your edited regions.

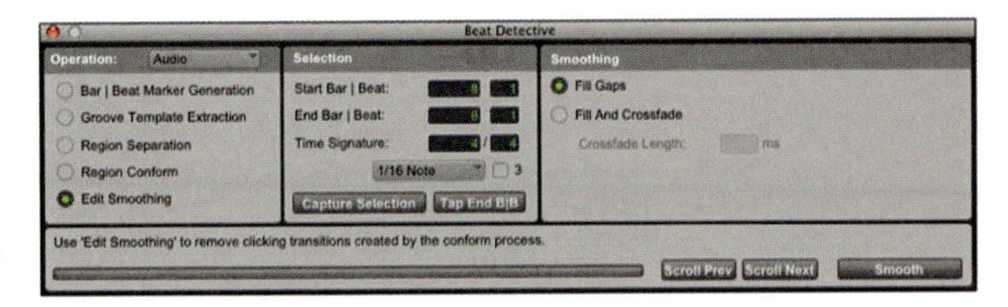

Figure 5.57 Using Edit Smoothing.

Figure 5.58 The Smoothing options of the Beat Detective window.

Tip

You can create one region out of many by using the Consolidate command found in the Edit menu. Simply select all regions (pretrack) that you want to consolidate to one region and apply the command.

When editing, you will have gaps between your work that will create pops and clicks in your recorded work. Gaps created between regions when conforming may be filled using the Edit Smoothing mode of Beat Detective. This process automatically fades and crossfades the ends of regions to prevent clicks and pops. To use Edit Smoothing mode, first select the regions in the Edit window and then select the appropriate Smoothing options from the Beat Detective window. The Fill Gaps option automatically trims region endpoints so that the gaps between regions are filled. The Fill Gaps and Crossfade trim regions and automatically adds a Pre Crossfade before each region startpoint. Click the Smooth button to smooth the edits contained in the selection. Smoothing options are shown in Figure 5.58.

To complete the Beat Detective process, consolidate regions by selecting the appropriate tracks and then choosing Consolidate from the Edit drop-down menu after using Edit Smoothing to permanently print and commit your fade information into the regions. The Consolidate command creates a single file from the multiple regions. Now you have completed processing audio with Beat Detective and, hopefully, if all went well, you should have a smooth playing, quantized, and tight-sounding performance and, if not, then you can fix it as needed. Beat Detective is hard to perfect because it truly relies on the performance – especially when recording audio such as that from an acoustic or electronic drum set.

If you recall earlier, we mentioned that you could also do the same with your MIDI tracks. In Figure 5.59, you can see that you can select a MIDI track in the Edit window and process it with Beat Detective the same as you can an audio track.

Figure 5.59 The MIDI Beat Detective window.

Figure 5.59 show only two options: Bar|Beat Marker Generation and Groove Template Extraction. They operate much the same way as they do for Audio, with obvious changes to reflect the use of MIDI data. You will not need to separate regions, conform them, or smooth them when working with MIDI. Chapter 3, Composing, covers recording and editing MIDI in more detail.

5.12 Summary

In this chapter, we covered the editing features of Pro Tools 8 LE. You have learned to use these powerful editing features to improve the sound of your recording project at this step of the workflow. We covered the use of the Edit window and how to configure it for an editing session. We also made more session adjustments and learned about getting around the myriad windows available to edit with. We learned more about regions and how to select and further process them in an editing session. The Zoomer, Grabber, Trimmer, Selector, and Smart tools were covered in detail, as well as many other features and tips to working efficiently.

We covered creating fades and crossfades, which are used to give your recording a high-quality sound and feel, and also learned how you can trim off silence and blend regions together so that everything sounds smooth. We worked with the Fades dialog, where you can create and manage your fades. We also learned about the different types of fades offered with Pro Tools 8 LE and how to create them. We also learned how to perfect your recordings with tools such as Beat Detective. In our next chapter, Mixing, we will learn how to take all your editing work and combine all of the voices (tracks) into a mix that sounds balanced as well as exciting. This takes a lot of listening so that you can get the best-sounding mix possible.

In this chapter

6 Mixing

In this chapter, we cover how to mix and prep your work for final mastering with Pro Tools 8. We cover mixing concepts, use of effects, automation, and much more. We also take a close look at not only using digital plug-ins but also outboard gear to enhance your mix and then your final mixdown.

6.1 Introduction

Now that you have completed your editing process, you are ready to further enhance your work by adding effects, automation, and setting a final mix! Getting a great mix is not easy and takes time and a lot of listening to your playback. When you have prepared a mixdown, you can export it (see Chapter 7) and finalize your work for delivery. So if you can, be patient, as there is still much to do before we have a finished product. You should, however, have a good picture of what your work will sound like once completed.

In the last chapter, we covered the editing process and how to polish your work. In this chapter, we are going to cover all the work that goes into making your polished work sound even better. We will cover the adding of effects and how to configure Pro Tools 8 LE for mixing the tracks, so your sound is balanced and sonically perfected. We will cover panning, monitoring, and routing, and will finish the chapter with a section on how to master your work and create a Red Book CD-ROM.

Before we get started with the actual mixing session within Pro Tools 8 LE, we need to control the environment in which the mixing will take place, select good monitor speakers, and place them correctly. Monitor selection and placement, and using a variety of monitor speakers, are key to successful mixing sessions. No matter how good your ear may be, not having the correct equipment, not knowing how to use it, or not having enough of it will set you back tremendously when mixing. For example, in Chapter 1, we covered the treatment of your workspace. By not applying a bass trap correctly, you may wind up mixing in a room that does not give you the proper mix. If your speakers were not placed correctly or if you are mixing in a particularly bass-laden area of the room, your mix ultimately could be thrown off, leading to a poor mix.

Prior to mixing, review your DAW design and setup to ensure that you have the correct tools needed to complete the job. Mixing requires the use of monitors and/or headphones for playback. Headphones are not recommended for mixing because they change the stereo imaging and often have an inaccurate frequency response. As mentioned in Chapter 1, the main setup of your DAW included playback monitors (or speakers) and/or a pair of stereo headphones (for overdubbing). In this section, we also cover the use of analog gear and summing. This opens the door to using many new effects with your DAW that add the color and warmth of analog to your effects arsenal.

Caution

Protecting your ears is paramount to being an audio engineer and producer. When mixing, what and how much of what you can hear is imperative to great mixing. You should always work to protect your hearing. This will help you to hear your mixes with more clarity and give your changes and adjustments more accuracy. Many times, musicians and engineers rely on "foamies," which are practically useless. They still allow dangerous levels of sound through to your ears, no matter how snug a fit you think you have. If you are working as an audio engineer or producer, it's recommended that you use an appropriate tool to protect your hearing. Figure 6.1 shows hearing protection that dramatically reduces the harmful effects of sound above 80 dB.

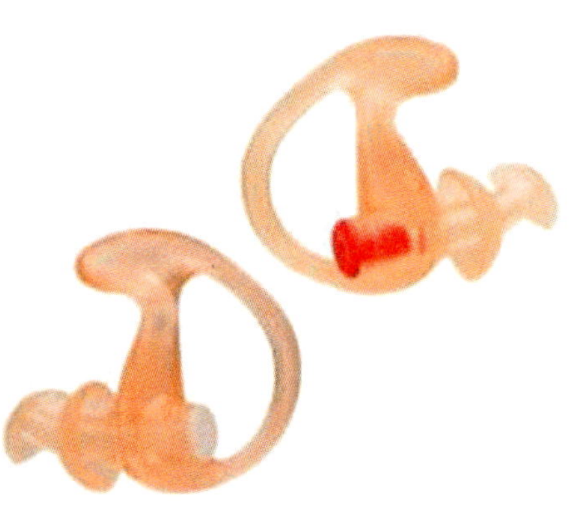

Figure 6.1 Using proper hearing protection.

These tools protect your hearing without interfering with your music. They are cost-effective and protect your hearing without interfering with your ability to play. There are many different types you can get. Make sure that you get ones that are created for protecting your hearing (http://www.hearos.com/).

When working with headphones, it's important to consider the quality of the drivers in the headphones and the overall comfort of the unit. If possible, find a place where you can test them first to see what kind of sound they reproduce. You will want to get a set that is comfortable, especially if you are using them as your only source for playback and mixing. Make sure they fit well because some headphones have a tighter fit, while others are less restrictive. Just because some cost more doesn't make them better than others, so take time to investigate and see which ones you like and want to work with.

Tip

If you only have the option to mix in your headphones (if you need to remain in a quiet room or if you cannot afford expensive powered speakers), you should have a few different sets of headphones to choose from. High-quality headphones may be costly, but they generally produce high-quality sound as well as provide comfort. When planning to mix, you should not solely rely on cheap headphones or Apple iPod earphones because they have poor bass response. You should, however, listen to your mix though a pair of cheap headphones to see what your mix will sound like on them.

Note

Don't mix on floor monitors (wedges) because they are seldom accurate – use studio monitor speakers instead. Floor monitors are often used in live sound situations and are used to keep a mix of what artists can hear while onstage. Digidesign VENUE systems are used for live sound reinforcement to meet this need.

Using professional monitor speakers is the next step in preparing for a mix. In Chapter 1, we discussed the importance of selecting and using monitors in your studio. If you have a professional studio and plan to deploy multiple sets of monitors for reference, there are many steps you can take such as building speaker soffits, as well as isolating them on your desk or individual stands. The importance of treating your workspace is critical when mixing because you may hear things that aren't in your mix at all coming from speaker vibrations coming up through the floor. Speakers that are not isolated or shock mounted can have inaccurate bass response. Speaker isolation is simple to do. Figure 6.2 shows a set of isolation wedges used for amplifiers, microphones, monitors, and stands. You can fabricate your own wedges using similar materials.

Figure 6.2 Isolation wedges.

If you are working in a small project studio with poor acoustics, use a set of near-field monitors on your desk or on stands to reduce the influence of room acoustics on the speakers' sound.

Note

Sit 3–4 feet from near-field speakers for best results.

You can also mount speakers into soffits in your wall.

It's recommended that you have multiple playback monitors to choose from when playing back your work so many times. Professional engineers will spend time mixing with headphones and with multiple sets of monitors (stereo and mono), as well as other playback systems such as car stereos, computer speakers, iPhones, and so on. If you can make the mix sound decent in all those formats, you have succeeded in creating a near-finished work. It may not sound exactly as you wanted it, but that's why the composing and recording phases are key to getting the sounds you want up front. The mixing phase only adds to them and balances the work as a whole to prepare it for final exporting and delivery.

Tips

When mixing, it's recommended that you get a small mono system for playback. This will give you the best representation of the worst system your music will be heard on. If you can mix it to sound great in a small mono system, mixing your work in a bigger system becomes even easier. Learn how to mix at very low volumes and higher volumes so that you can hear your mix as it will be heard by those listening to it on many different systems at many different volume levels.

For project studios, you need a set of stereo headphones, a set of near-field monitors, and a good stereo speaker set. Add in a cheap mono playback system and mix at very low volume and you will have the ability to get a decent mix in almost any room. Many engineers recommend a consistent mixing volume level of 85 dB SPL. If you mix quieter than that, your ears become less sensitive to low frequencies, so you may mix in too much bass. If you listen louder than that, your ears become more sensitive to upper-mid frequencies, so you may not have added enough Equalization (EQ) in the upper midrange.

Sit exactly between your monitor speakers, as far from them as they are spaced apart. Place them at ear height. You do not want to place your monitors on too steep of an angle either. Some foam isolation kits come preshaped and angle the speakers for you, so you should be aware of your ear height while designing your layout. Monitors have to be positioned perfectly. It's recommended that you place your speakers first, and once you adjust the height, then adjust the angle for optimal listening.

Tip

When purchasing monitors, you need to take the size of your room into consideration. You do not want to spend a lot of money on monitors that are too powerful for your room. Most near-field monitors sold today are designed to work well in smaller rooms.

Some monitors are sold with subwoofers. You can choose this option if you want to supplement the low-frequency response of small "satellite" speakers.

Note

The room you mix in can also be further treated with diffusers that will help to keep sound from echoing through a small room. Foam above your head on low ceilings and bass traps in the corners of your room to reduce standing waves are helpful. Low frequencies do not "see" parallel or non-parallel walls – that applies only to midfrequency and high-frequency sounds. Bass traps absorb low frequencies. Putting them across room corners is effective because bass builds up, especially in corners.

Now open up a session that you have edited and prepare it for the mixing phase of your workflow. You can configure the preferences and get started with your mixing tasks.

6.2 The Mix window

The mixing phase of the workflow is primarily tackled within the Mix window. The Mix window can be found within an opened session by going to the Window menu and selecting Mix. The Mix window is displayed in Figure 6.3.

Figure 6.3 The Mix window.

Notes

As you can see from the Mix window, all changes have to be made with a keyboard and/or mouse. For that reason, you may find that mixing and configuring automation are easier to do with a control surface. Control surface selection and configuration was covered in Chapter 1 and its configuration within Pro Tools was covered in Chapter 2.

Many professional engineers and producers refuse to "mix inside the box," which is to mix with effects entirely in the computer and save the mix to the computer's hard drive. Therefore, before you start your mixing session in Pro Tools, be aware that the end of this chapter covers the use of a master deck, where you can also record your mix instead of on your hard drive. Be aware of this option, as many prefer to use equipment with higher sample rate options and better toolsets, among other benefits.

Let's get Pro Tools 8 LE ready for your mixing session. We will need to add plug-ins, configure automation, and prepare the work for final mastering. To get your session ready, simply open up the Pro Tools Preferences found in the Setup menu. You will want to adjust your Playback Engine again. As we learned in Session Setup, you will need to increase your H/W Buffer Size (as seen in Fig. 6.4) so that you can run extra plug-ins and work while making changes. If you do not, then it's likely you will get dropouts, clicks, or computer crashes.

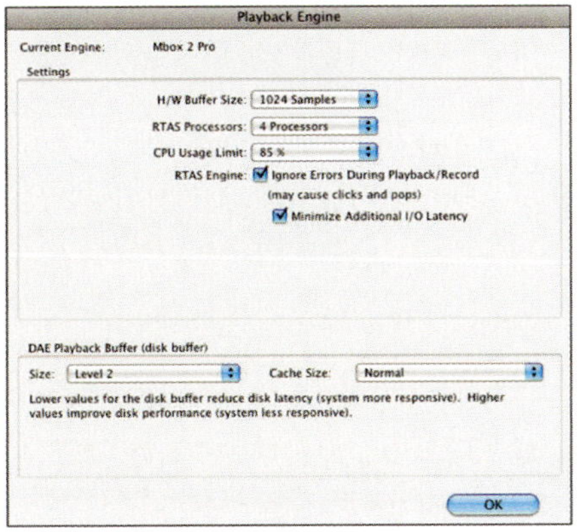

Figure 6.4 Setting the Playback Engine for mixing.

Open a session that you want to mix.

You will need to understand tracks to mix and add automation or effects. As we learned in Chapter 2, the track components from top to bottom when viewed in the Mix window are Inserts, Sends, I/O, Automation, panning (if stereo), record, a fader, indicators, comments, and identification. A stereo channel strip that controls one stereo track can be seen in Figure 6.5.

Figure 6.5 A stereo channel strip in the Mix window.

When mixing, you will work with multiple track types depending on the size of your production. If you are working with a full band, you may have many tracks to mix and configure effects. You can mix them all simultaneously in the Mix window (Fig. 6.6). They are denoted by track color as well as by the symbol located to the top right directly above the track name field.

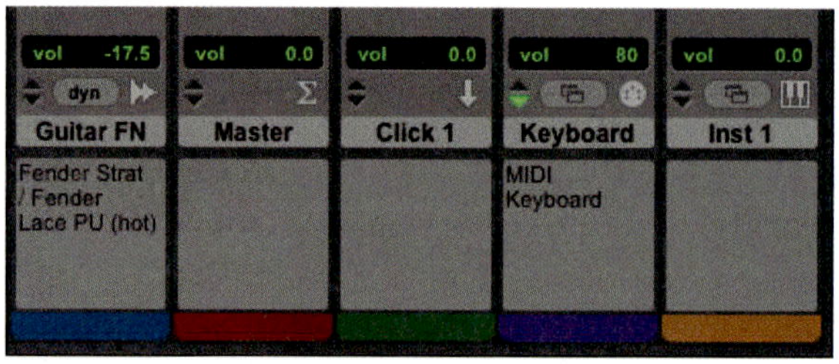

Figure 6.6 Viewing all track types in the Mix window.

Tip

If you click on the track type symbol and hold down your mouse button, you can render a track inactive. This is helpful if you do not want to delete a track but want to remove it from the track count. For example, if you are running low on tracks, Pro Tools will not allow you simply to create any more. If you render a track inactive, you can create a new track. If you reactivate it, it will activate but not be usable until you free up another track.

Your main tracks – Audio, Master, Auxiliary Input, MIDI, and Instrument – are easily seen and identified within the Mix window. The process of mixing is not difficult to do, but it will be time consuming depending on the project you are working on. For example, with the Production Toolkit, you can expand to 96 tracks. Imagine trying to balance each individual track so that it can be bounced down to a stereo mix! This is why the Mix window is handy as its own stand-alone window. As mentioned in Chapter 5, Editing, you can customize the Edit window to also incorporate the most of the same options seen in the Mix window. Figure 6.7 shows a track in the Edit window where you can adjust settings and add effects similar to the Mix window.

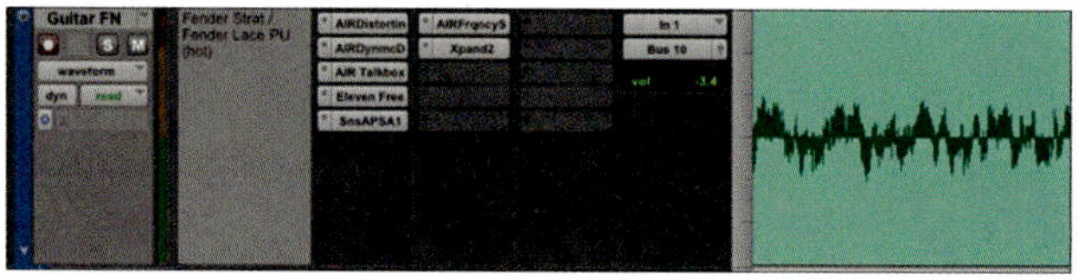

Figure 6.7 Viewing a Track with Mix options in the Edit Window.

In this chapter, we focus primarily on the Mix window and show what can be done with it. When looking at the Mix window when you have a lot of tracks to work on, it can seem to be a formidable task to organize them and get them balanced. This is why it's important to have multiple mix sessions because you will likely use the first one to get a baseline and then improve on it from there. Of course, this is if you are recording 12 separate drum tracks, five instruments, and a vocal track. If you are mixing a radio show or podcast, it's likely you can get a great mix within a few short adjustment and playback sessions.

To mix well, you really have to approach the project in sections. First, try to listen to the work as you recorded it and make obvious changes. For example, if the snare drum is extremely prominent in the mix, you can turn it down a little. If the drum's cymbals are washing the vocals out, turn them down a bit. Get a basic fader level for the entire mix and ensure that you are not boosting the gain of any fader above 0 dB or making any track clip. A good example of getting a proper level is seen in Figure 6.8. Here, the signal is strong (high) and therefore sounds powerful without distorting (clipping).

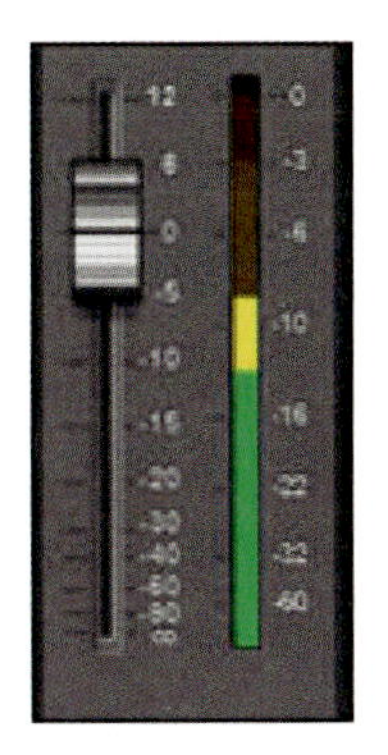

Figure 6.8 Example of a proper level.

Now that we know our way around the Mix window a little better and know what we will do within it, let's open a song or project and add effects. In the mixing process, adding effects often enhances the sound of individual instruments or voices or even the whole recording.

Note

With Pro Tools 8 LE, you now have the option to show portions of your track below the main track on their own subtracks. Controls (like panning, volume, etc.) can be adjusted in the Edit window while editing and mixing. As we will cover later in the chapter, automation can also be used here as well.

6.3 Using effects

In the world of DAW mixing, effects can make or break a recording. In this part of the workflow, your recording should be edited and you should have a basic mix. Now, you can continue your mixing process based on your initial assessment and also determine whether anything needs further processing.

Caution

Overprocessing your work generally causes a degradation of the original signal. Try to get the best quality sound you can before fixing anything in the mixing process, as this process is used to enhance the sound, not fix it.

Starting in Chapter 1, we learned about I/O and how to apply it while recording. Here in the mixing stage, the routing is the same, except we are going to use it to save processing power. You can easily create an Auxiliary Input track and insert the effect plug-in in it, then route (bus or submix) other tracks through it. This essentially runs the plug-in one time, therefore saving resources by avoiding running the plug-in multiple times on individual tracks. This setup can be used to apply one instance of compression to several background vocal tracks. To configure effects, you can create a new Auxiliary Input track and add an effect to it via the Inserts section. Then you route multiple tracks to the same bus number that the Auxiliary Input track is using. Figure 6.9 shows the new I/O section configured to use Bus 10.

Figure 6.9 Routing multiple tracks to an Auxiliary Input track.

Bus 10 is used as the routing path to get all your tracks to the Auxiliary Input track. The Auxiliary Input track is then configured to send the mix to Analogs 1 and 2 (the main stereo outputs of your audio interface). Next, all you need to do is configure the plug-in(s) on the Auxiliary Input track and then route your track outputs to the same bus number that the Auxiliary Input track is using. Figure 6.10 shows the routing of one track to Bus 10 for further processing.

Figure 6.10 Routing a track to an Auxiliary Input track.

Next, you can check fader levels on each track's output, the Auxiliary Input track, and finally the Master Fader (if used).

When applying effects, it's important to know how to add them through the Insert section. Simply click on a dark rectangle in the Insert section, select plug-in and then choose which one you would like to use (Fig. 6.11).

Figure 6.11 Adding a plug-in within the Mix window.

Note

Remember that plug-ins may or may not work on specific tracks depending on track type. For example, some stereo plug-ins will not load in a mono audio track. Multi-mono and Multichannel plug-ins are used on stereo tracks and have slightly different features than their mono counterparts. Usually, these plug-ins are specifically designed and will not interoperate between mono and stereo tracks. If you try to move or copy a plug-in from a stereo track to a mono track, Pro Tools will not allow you to move it and will give you an error message. Multi-mono plug-ins are used when you are using plug-ins that do not require the signal to be on discrete channels, which is not the case with Multichannel plug-ins. Multichannel plug-ins are used when true Multichannel DSP engines are used.

Note

Multichannel audio tracks are used to create 5.1 surround sound mixes from a single track.

To add a plug-in, click on the Inserts selector and click on plug-in, then select the plug-in you wish to use. In Figure 6.12, you can see the addition of the new AIR Kill EQ, part of the creative pack that comes with Pro Tools 8. This tool is used primarily for remixing and complete removing of specific frequencies in the EQ range and can be very helpful in tightening up your mix.

Figure 6.12 Adding the AIR Kill EQ plug-in.

By adding effects, you can now process tracks and do a before-and-after comparison. You can bypass them as well and go back to the original sound without removing them. You can also drag and drop them or copy them from one location to another. A copy can be done by holding down the Option key on your keyboard and simply dragging a plug-in from one track to another.

Note

New AIR Instrument Plug-ins include Boom, DB-33, Mini Grand, Structure Free, Vacuum, Xpand!2 as well as many others including Distortion, Dynamic Delay, Ensemble, Filter-Gate-Sequencer, Flanger, Frequency Shifter, KillEQ, Phaser, Reverb, Talkbox, and Vintage Filter. Other new tools include Eleven Free and the Bomb Factory/Tech21 Sans Amp, which were covered in Chapter 3, Recording.

Note

Effects are processed in order from the top of the track's insert path to the bottom (A–J).

Tip

To save time, make a template. You can create all of your Auxiliary Input tracks and I/O maps in a session called "mixing" and load up some of the most common settings you use from session to session. This will save you time when creating a session.

You can apply effects in the mix and in the mastering phase. You will apply many effects to your work during mixing that you may also apply in the mastering phase. You should avoid overprocessing because if you, for example, overdo compression in the mix, you cannot remove it during mastering. This is why it's important to not overprocess your work. Also, record mixes at about –3 dBFS maximum to leave some "headroom" in your mix to allow for extra processing in the mastering phase.

Note

Mastering in the professional studio environment is normally done by a mastering engineer in a mastering house. This just means that a second set of ears approaches your mix with new ideas, and the engineer is likely a specialist in mastering and has the proper equipment to get you a Red Book–quality CD.

EQ, compression, reverb, and delay are just some of the effects you can use to enhance and accentuate your tracks or overall mix. To understand each, we need to look at what you will do with each in a production cycle. Your most important job when mixing is balancing all your tracks and using EQ. This device or plug-in will help you to select a range of audio frequencies and turn them up or down. For example, use a 100 Hz high-pass filter (low-cut filter) on electric guitar so that it doesn't interfere with the bass guitar sound. You can use the EQ to filter out highs and lows in your tracks so that only the "sweet spot" of each track is heard. This unclutters the mix and gives your mix clarity, punch, and definition.

Note

> The Q, which is the range of frequencies that an equalizer affects, is set with a parametric EQ.

When working with compression, you need to remember that you shouldn't compress without a plan. Compression applied incorrectly can really destroy a great recording when it comes time to mix and master. Figure 6.13 shows an API 2500 Compressor, which is often used to squeeze a guitar track to give it more punch. Here, you can make multiple adjustments to your settings, such as setting the threshold at –12 dB for best results.

Figure 6.13 Adding compression to an electric guitar track.

Compression will take the track's instrument and squeeze it, making it sound punchier and louder, and also possibly more distorted. Compression will be covered in more detail when we discuss the mastering phase at the end of this chapter.

Reverb can be used to give the recorded tracks or overall mix a "live" sound. For example, if you wanted to make an acoustic guitar sound more ambient, you can make it sound as if it's recorded in a large room as seen in Figure 6.14. Here, we add a simple Reverb plug-in and set its presets.

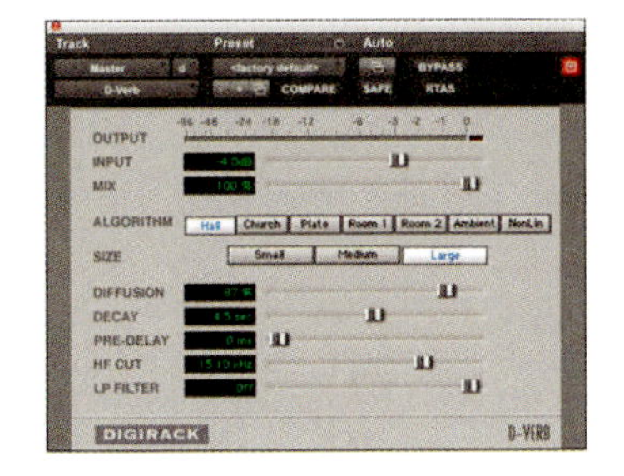

Figure 6.14 Adding reverb to an acoustic guitar track.

You can modify the presets in various ways. For example, if you wanted to give drums a bigger room feel, you can add a small reverb preset and shorten the decay to one second or less for punch. Other options include using a short-plate reverb with a decay of 1–2 s mixed with an 80 Hz HP filter to remove low-frequency "mud" in the reverb. As you can see, EQ, compression, reverb, and multiple effects can really bring out the instruments and vocals in your mix and make a small-room recording sound like it was recorded in an amphitheatre. Other tools to add effects include Chorus and Flanging. A Chorus is sometimes used to build up a sound much like a vocal choir does to a passage of a song. A harmonizer will work within a key's octave range and add layers to it so that a note on a keyboard is fattened to sound like multiple pianos playing at once in different intervals of the same key. You can also experiment with different settings within reverb as well as chorus, flanging, and delay. Figure 6.15 shows the use of a delay (echo) effect on the Master Fader bus. You would turn up the aux sends only for those instruments you want echo on. Normally, effects are inserted in an effects bus (Auxiliary Input track) rather than in the Master Fader (stereo output) bus.

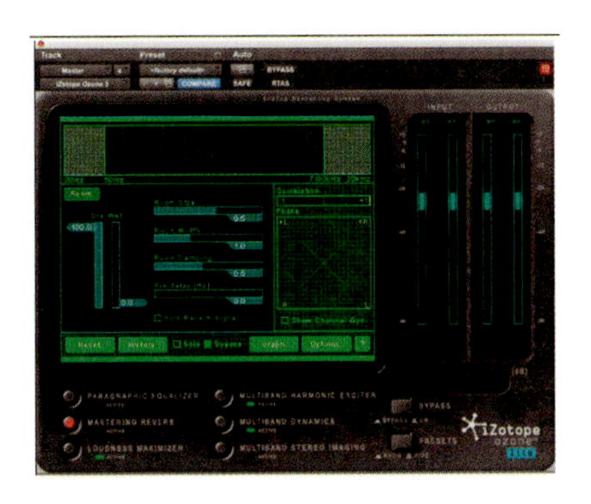

Figure 6.15 Adding delay to the Master Fader.

You can see the adjustments you can make to the delay to shorten it, lengthen it, adjust the sound to be wetter or drier, and so on.

Vocals tend to take a lot of work when you begin mixing and adding effects. This is the one part of mixing that is not easy to achieve if the performance is weak. This is why Autotune or other pitch-correction tools are often used on vocal tracks to correct the pitch of flat or sharp notes.

Once you have added and experimented with plug-ins, your next step is to understand the use of AudioSuite. Found in the Edit window, AudioSuite is a subset of the same plug-in tools you just used while in the Mix window. Some of these tools can be used in both windows, whereas others can only be used in one. For example, if you wanted to take a cymbal hit recorded on an audio track and "reverse" it, you could find the region in the Edit window, select it with the Selector tool and then go to the AudioSuite drop-down menu, go to the subsection of effects you want to choose from (such as Other) and select the tool you want to use. If you choose Reverse, you simply need to click on Process and the selection will be reversed. To get to AudioSuite, simply go to the AudioSuite menu and select the tool you need (Fig. 6.16). When working with Pro Tools 8 LE, you can use AudioSuite to apply effects to regions.

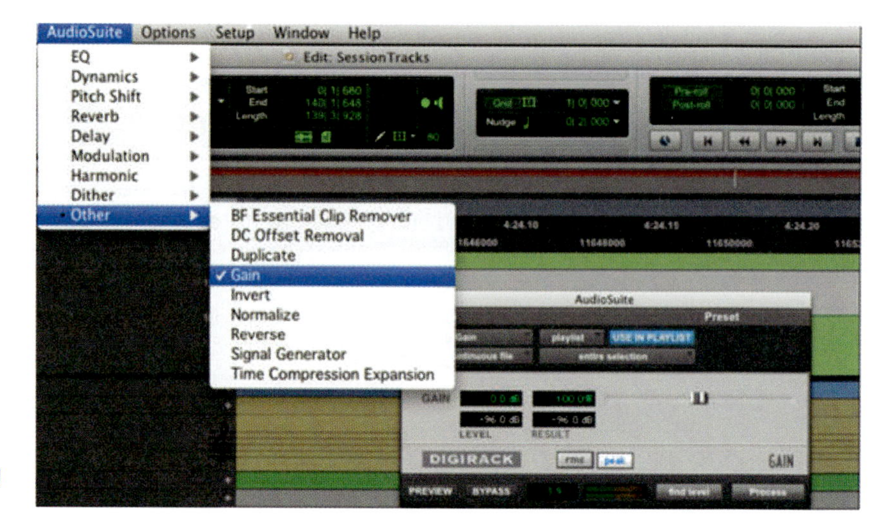

Figure 6.16 Working with AudioSuite.

Here, Gain is selected to add more amplification (and thus distortion) to the track. Maybe the signal is too low and you want to boost it manually. Select the section you want to adjust if you have not done so already, select the settings you want to change, such as increasing the gain, and then press Process, as seen in Figure 6.17.

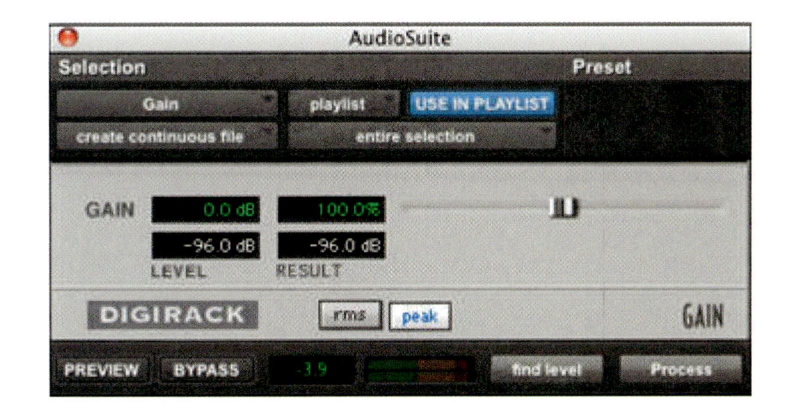

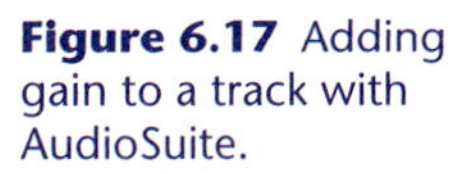

Figure 6.17 Adding gain to a track with AudioSuite.

As mentioned, you will not be able to use some of these tools (such as Reverse and Gain) in the Mix window, so remember that if you want to use the effects found in AudioSuite, you need to work within the Edit window.

6.4 Recorded drums

If you want to do a live drum/percussion setup with a keyboard, guitar, bass, and vocal, here are some general recommendations for configuring your mix. For drums, you can specifically enhance the drums by using gates, especially when placing a microphone inside a kick drum. You should be concerned about getting tone through EQ techniques and attack through compression. Ambience is important so that your drums do not sound flat and lifeless, and a small reverb setting is useful in achieving it. The snare drum is also critical to every mix, especially on live drums. All hits must be consistent and hopefully Beat Detective (covered in Chapter 5, Editing) was used to tighten it up. To bring out the snare in the mix, you can give it a little reverb and make sure that it's not mixed too loudly and competing with the vocal. High end should be filtered out on both the drums and the vocals to ensure that they do not step on each other and can both be heard clearly. Overhead microphones on the drum kit provide extra ambiance as they are able to capture the entire sound of the top of the kit. If spaced and set up correctly, they can add a lot of top end to your recorded drums. Make sure you do not pan them too wide (which is considered the spatial position the track occupies between your monitor speakers) or too far to the left and right.

Cymbals can be tricky if they are not recorded correctly. It's common to apply a 500 Hz high-pass (low-cut) filter to the overhead mics so that they pick up mainly the cymbals and not so much of the toms. Figure 6.18 shows the configuration of a SSL G-Channel (Waves) Hi Hat EQ filter used to carve out the frequencies and filter out the rest.

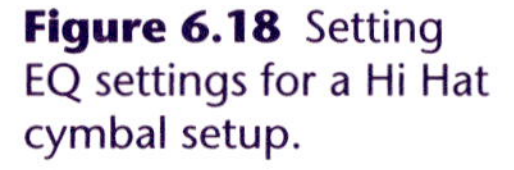

Figure 6.18 Setting EQ settings for a Hi Hat cymbal setup.

Generally, to reduce low-frequency leakage into the cymbal mics, you need to also filter out the lows with a high-pass filter.

6.5 Equalization

To work with an EQ, you should know what its settings are. Table 6.1 shows some basic and common settings you can configure when working with an EQ and recording a live band (as an example). Here, you can configure the most generic frequencies you would "carve" out for your recording.

These settings can give you a head start to cleaning up your mix. By making these adjustments to each instrument, you can increase the clarity. Although these are just recommended settings that many engineers have used over the years, they really do give you a great starting point to beginning your initial filtering, equalization, and final processing work. These settings are also easy

Table 6.1 Common EQ settings for live instruments

Drums	−4 @ 400 Hz +4 @ 15 kHz
Guitar	−4 @ 100 Hz +2 @ 3 kHz
Bass	−2 @ 50 Hz +4 @ 400 Hz +2 @ 1.5 kHz
Keyboard	+4 @ 5 kHz
Vocals	+4 @ 10 kHz +2 @ 5 kHz @ 200 Hz

to key in. Boost/cut (gain) is measured in decibels and the EQ frequency is measured in Hz. Of course, the EQ settings you actually use depend on the frequency response of the mics and their placement, and on the sound of the instrument itself.

Tip

Use an EQ to reduce the masking effect. When two instruments occupy the same part of the frequency spectrum, they can mask or hide each other. This is why the bass guitar often disappears in a mix – it is covered up by low frequencies in the kick drum or guitars. You might remove some low frequencies in the kick or guitars so that they don't mask the bass guitar.

Table 6.1 shows some common EQ settings. You can use these settings on any EQ that has the frequency bands to allow for it. Some EQs (such as parametric EQs) also have more capabilities to fine-tune your recording and make it sound even better, including the setting of "Q" or bandwidth. The Q can be set on any Parametric EQ that has that function. There are many different EQs and each type has its own operating procedures, so it's recommended that if you work with EQs, you spend time to learning and working with them and experimenting with your own settings.

Note

High-Q settings affect a narrow band of frequencies (such as one-half octave); low-Q settings affect a wide band of frequencies (such as two octaves). The amount of boosting and cutting (reduction) at a particular frequency is measured in dB.

When recording, you need to consider every instrument and where it will sit in the sound spectrum. The bass (or bass guitar) maintains the lower frequencies of the spectrum. It's important to control all other instruments and high-pass filter them so that you can hear the bass without having to crank it in the mix and distort it. Many times bass is recorded with a DI box, other times, a recorded amplifier with a cabinet and microphone is used. Either way, the same concepts for recording guitar apply for recording bass – except they require different EQ settings. Usually you record flat (without EQ), then apply EQ during the mix.

6.6 Panning

Panning lets you adjust the left-center-right position of a track between your monitor speakers. For example, for a two-guitar band, you can pan one guitar left and the other guitar right. Whereas in reality, the tracks are heard in both ears, panning places an instrument to one side over the other and helps the listener distinguish two similar instruments.

Panning is done for every instrument and voice, including overhead cymbal mics that are more likely to be a stereo pair – so you can pan one microphone left and the other right in the mix. The drums are usually spaced exactly how you would see a typical five-piece drum set. You can put the snare slightly off-center with the kick and place each tom how you would see it located from left to right. You can pan the floor tom more to the left and you can place the Hi Hat more to the right. You can send the vocals straight down the center of the mix. Any other instruments obviously can be left of the center or in the center itself. It's okay to try different things and experiment with panning techniques.

Note

Panning is done left to right. If surround sound is used, then you can pan left to right, as well as back to front.

Tip

Consider frequency response when panning. High-frequency instruments are directional, whereas low-frequency instruments are not. It's recommended that you put the lower register tracks toward the center of your stereo spectrum. For example, the bass guitar, bass drum kick, and toms should usually stay closer to the center of your mix. High-frequency instruments such as a tambourine should be panned left and right.

6.7 Layering

Layering or doubling tracks is helpful to get a very thick tone and "fatten" up the sound. It's recommended that you use multiple instruments with multiple amplifiers using different cabinets and different microphones and that you test the amounts of midrange and distortion needed to achieve your sounds. You can sometimes layer your guitar simply by recording it with a plug-in, duplicating it, and then changing the plug-in sound. This simulates the process that we just mentioned, but it's all done by preference. Some professional engineers refuse to use plug-ins for recording certain things, such as guitar, whereas some swear by the fact you do not need an amplifier anymore with the right DAW setup. It's a good rule of thumb that you should always go with what you like best and try a mix of both. Usually, a mix of both produces the sound you need.

Note

When mixing and applying sounds, you will find that less is sometimes more and vice versa. When new to recording, you may think that tons of saturation, distortion, and gain are needed to achieve a sound but later find out that sonically it creates one you didn't expect. For guitars especially, easing off the gain in the guitar-amp emulator plug-in a little can produce a thicker sound. It's also recommended that if you use hot pickups that produce a lot of feedback, you also use a noise gate to reduce noise between musical phrases.

6.8 Grouping tracks

To group tracks is to assign several tracks to a bus or stereo channel so that you can control all the tracks at once. For example, you can fade two keyboard tracks at once if they are in the same group. Simply group the tracks by going to the Groups section of the Mix window and dropping down into the menu from the drop-down arrow. Here, you can create a New Group, Display your groups (Edit or Mix groups), as well as Suspend All Groups, which temporarily disables them. You can see this menu setting in Figure 6.19.

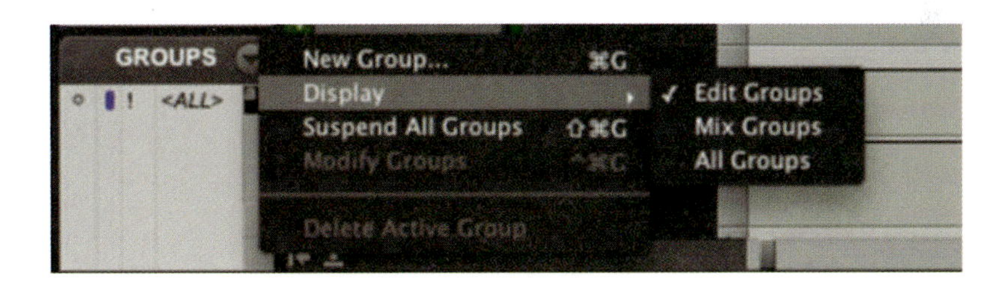

Figure 6.19 Group options.

In the Groups section, you can modify, edit, and delete your groups. Once you have created a group, you will be asked to name it and add tracks to it. Figure 6.20 shows the Create Group dialog.

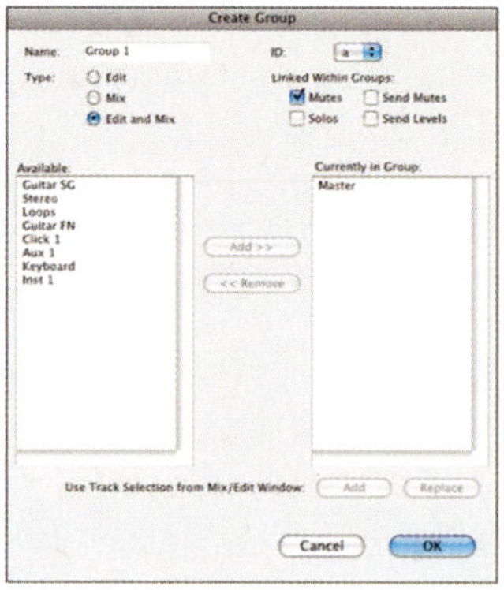

Figure 6.20 Create Group dialog.

Here, you can add your guitar tracks, for example, and mix (or edit) them together. This will undoubtedly save you time while working with many tracks.

6.9 Using automation

Automation is the process of recording (writing) and playing (reading) control changes for a track. Your computer remembers your mix moves, then recreates them whenever you play the multitrack project. For example, if you wanted to apply a volume change to a track such as fading out a song, you can do that manually with your control surface and have the automation data written to be played back (read) when the selection is played back. You can automate (write) fader moves, effects-send levels, panning changes, and so on. These changes keep the mix interesting and dynamic. To add automation, simply open the Edit or Mix window and click on the automation section as seen in Figure 6.21. Here, you can see the first track to the left of the Mix window where automation is being applied.

Figure 6.21 Adding automation to tracks in the Mix window.

First, you select what you want to automate. You can do this by going to the Window menu and selecting the Automation window. This window will give you a set of options to choose from, such as panning and volume. Figure 6.22 shows the Automation window.

Figure 6.22 Adding automation tasks with the automation window.

Once you set what you want to edit (volume as an example), you will need to configure the track mode by arming the track to "write" and then arming the track to record. For example, to turn automation on in "write" mode, select "write" from the Automation menu (the other options are read, touch, and latch mode). Once you set it to "write," you can then click on the record button. Figure 6.23 shows a track's Automation section configured to write automation.

Figure 6.23 Configuring the track to write automation.

Next, you can use your control surface or Mix window faders to decrease or increase the fader volume. Once you have finished, disarm the track and turn off automation. You can apply the same steps to adjust panning, muting, and so on. To record using any other mode, such as touch or latch, simply set the mode and follow the same steps to record. Touch mode is used to write automation data where the position of the fader is memorized so that when you adjust it, it will snap back to the original position when you release it. Latch mode is used the same way Touch mode is used, expect that Latch mode will continue to write the recorded automation until you hit the stop button on the recorder. Both are handy in making specific edits such as quick volume fades and so on.

Tip

You can view automation easily by using a separate playlist created in the Edit window. Change the track view to see it or set up an automation lane underneath the main track.

Note

Subtracks found below main tracks in the Edit window will contain a view of your automation changes, for example, you can see the volume changes made when you faded them. These tracks contain controller data used to further edit your work and are helpful for making detailed edits on automated and nonautomated work.

6.10 Final mixdown and mastering

Once you have completed the mixing process by balancing your tracks, adding effects, panning, adjusting volume, adding fades and other effects changes with the automation process, and performing final processing on your work, you are ready for the final mixdown. The final mixdown is the step when you export the final mix before mastering. In our next section, we look at mastering and analog processing techniques. At this point, you will do something with the final mix, the one that you and the producer and musicians agree is the best.

If you would like to bounce the file, then this chapter and the next, Chapter 7, Importing and Delivery, will cover all of the steps that you need to take to prepare the file for a CD. If you would like to send the file to through analog gear into a mastering deck and create a Red Book CD, read on. These are the two most commonly used scenarios with today's DAWs.

When bouncing down a final mix, you should make sure that you check all of your Mix window settings. Ensure that all solo and muted tracks are set correctly and make sure that all tracks are not armed. You will not be able to bounce a session with a track armed.

To bounce a final mix to disk, all you need to do is quickly check your levels and invoke the bounce command by going to the File menu, selecting Bounce to, and then Disk. You can now export the file to your hard disk. Figure 6.24 shows the Bounce dialog.

Figure 6.24 The Bounce dialog.

From here, all you need to do is configure settings (covered in Chapter 7, Importing and Delivery), select a destination folder for the file, enter a name, and click Save.

If you want to do further mastering before you export, then you can route your final mixdown through a Master Fader. This would be as simple as creating the track and routing all your tracks to it. Then it can be used for final processing. All a Master Fader is used for is final DSP processing (across the entire mix), master automation (such as volume control), and ultimate fader control of the entire mix.

Tip

Use dithering on your Master Fader. Dither is a shaping tool that helps provide a better sampling experience, thus providing for better sound quality. A dither plug-in can be added to your final mix so that you can further adjust your work to the correct bit depth (16 bit is used for a CD) and make more noise-shaping choices. Dither should be used only when exporting a mix to a lower bit depth, for example, exporting a 24-bit mix to a 16-bit mix for use on a CD. If you are going to master the song or several songs, export it in 24-bit format, do the mastering processing in 24 bits, then turn on dither and export the mastered file in 16 bits, so it will play on a CD. Dithering should be the very last step before you create the file(s) that will be on a CD.

Note

The process of mastering should be the final DSP processing you do to your work. As mentioned earlier in this chapter, it's recommended that if you do not have the skill or equipment to do mastering, you send your work out to a mastering house for final processing. If you are doing the mastering in-house, let's learn how to get it done correctly.

You need to set up your final mix, save the project, and export (bounce) the mix to your hard drive. You then send that mix to a mastering deck (or import it into a mastering project on your computer) and "finalize" it. You can choose from many different types, but a commonly used (and recommended) cost-effective deck is an Alesis MultiLink (Fig. 6.25).

Figure 6.25 Alesis MultiLink mastering deck.

Tip

Use the Alesis MasterLink to do final equalization, compression, and normalization of your work. This unit can help to fine-tune your work into a "mastered" Red Book CD.

Now you can further process your work or burn it to CD with Red Book format. The Red Book standard was forged in 1980. It is a standard specification for the audio compact disc (CD-Audio) developed by Sony and Philips. All CD masters are Red Book format. The Red Book standard does not specify sound quality; it specifies the track start IDs, maximum level (0 dBFS = 16 bits), sample rate (44.1 kHz), and the physical configuration of the CD.

Many engineers and producers who own a mastering deck most times bypass adding most of their effects in Pro Tools and just use the mastering toolset that comes with the mastering deck. The Alesis MultiLink comes with many tools that allow you to further equalize, compress, normalize, limit, and dither your final mixdown. Mastering decks normally balance the tone with full-frequency spectrum equalization options, Q settings for tweaking, and multi-band compression for dynamic control.

Tip

When using the Alesis MasterLink, you can configure song fades. Fading out songs, or crossfading between songs, is very important to creating a professional CD-quality master, so make sure to apply your fades.

To send your final mix from your DAW to your mastering deck (such as an Alesis MasterLink) in stereo, you will need to connect your deck to the outputs of your Digidesign hardware with high-quality cabling, arm the mastering deck, and record in from Pro Tools. There is one more thing you can do to spice up your mix, that is, sending your mix through analog gear before it gets to the mastering deck for final processing. In the last section of this chapter, we cover how to send your final mixdown through analog gear to capture the warmth and color analog brings to a digital signal.

6.11 Analog processing

A common problem associated with DAW recording is the plastic sound or lack of warmth associated with many digital-based productions. This can be solved by using analog components that add character and depth to your sound. Since it is extremely difficult to emulate analog sound using digital technology or afford the steep costs associated with using this gear, it's no wonder many are taking a new approach to solving this problem. In this section, we cover how to inject analog sound into your digital mixdown before final mastering.

When adding analog components to your DAW, consider using an analog mic preamplifier and equalizer. In Chapter 1, Introduction, we discussed outboard gear and how to patch your outputs through this gear before connecting to a mastering deck.

Note

You can use a single Neve channel to expand your system. If you cannot afford an analog console and want to use a single channel, you can get a remake. A great solution to overcoming the huge price tag associated with a lot of high-end vintage gear is to try a remake or copy. This will accommodate the budgets of most who are trying to stay lean and also present a new benefit to the purchaser – a way to purchase equipment sold and shipped as new that produces the vintage sound many are looking to capture in their recorded works without dealing with the damage that sometimes comes with using ancient gear. Many analog preamps offer extra midrange EQ options that might be essential for both the recording and mixing stages.

To send your final mix from your DAW to your mastering deck (such as an Alesis MasterLink) in stereo, simply connect all your hardware and cables correctly and record your mixdown to your mastering deck. Before recording, simply play the track while testing the analog preamp. You can get your EQ settings and then copy the entire work, or a submix. When doing so, make sure that you do not clip the signal.

Tip

A Submix (or Stem Mix) is created when you send a submix or group mix signal to an analog component, as in taking all the drum tracks as a whole for example. You can send your submix through an analog compressor without affecting other tracks in your mix.

A great tool for analog compressing is an analog compressor and/or limiter if your budget will allow for it. This is the perfect complement to the analog preamp. Now you can have the flexibility to add analog compression to any recorded element, your submix, or your final mix sent to a mastering deck. Vintech Audio puts out an extremely cost-effective remake of the classic Neve 33609. The Vintech 609CA is shown in Figure 6.26.

Figure 6.26 Using an analog compressor for your drum submix.

This unit is a class A transformer-balanced stereo compressor and limiter. Using compression and limiting is not only important in the recording process, mixing, and then final mastering – it is essential to producing (or overproducing) your work. The 609CA has all discrete output amplifiers as well. You can send your signal from your DAW through the analog compressor and into the mastering deck. This will give you the best sequence for producing your sound after final mixdown. When you add analog compression, limiting, filtering, and equalization to remove the digital feel of your DAW, you open up a new toolset to work with. Although there are many who debate analog versus digital, there is nothing really left to debate. With sampling rates rising and the digital toolset getting deeper, it is getting better and easier to work with. Analog, however, does offers new sound quality and options and most rely on it to really produce the effects they are looking for.

A new world of analog effects are now open for your use, the only recommendation is that since you are adjusting the entire mix, it's imperative that you carefully monitor each mix so that you know how this equipment alters it. It's a common practice (although not mandatory) to put compression at the first step of the audio chain before final EQ adjustments, so you may want to consider running your DAW outputs to your compressor first, then finally to your mastering deck. This will allow for many new options and create some added flexibility for altering the sound of your final mixdown. Since audio engineering is part science and part experimentation, you should try different things such as adjusting the order of your audio chain to see how it affects your final mix.

If you choose to mix digital plug-ins with analog components, you will have many options with which to work. There are some amazing plug-ins that really add a lot to your recordings but most productions do not like to remain 100% digital as many feel it creates a sterile or lifeless overall sound. This doesn't mean that you should not use them or that they provide no real value – quite the contrary. When digital plug-ins are used in conjunction with some well-selected analog equipment, the mixtures can often produce some stellar results.

If a plug-in is selected for use, you can always use digital emulations of famous analog equipment such as Neve, API, or SSL from Waves. Waves is a company well known for their quality plug-ins that many would argue come extremely close to emulating vintage Neve, API, and SSL console sounds. With the release of their SSL-inspired line of software, you can use of amazing SSL filters to tighten up your sound. Figure 6.27 shows The SSL G-Equalizer from Waves.

Figure 6.27 The SSL G-Equalizer from Waves.

This four-band equalizer, modeled on the SSL G Series EQ292, is very helpful in giving your productions that vintage "SSL" sound. However, as mentioned before, to get the true sound of analog in your production, it's recommended that you infuse analog if you can. Digital plug-ins are great tools, but they do not quite match their analog counterparts yet. Using analog components (especially in the mixdown) is a great way to save money, get the analog sound you desire in a DAW and work with a new component that will help you with your recordings and mixing.

Now that you have finished your mixdown, it's time for the delivery process.

6.12 Summary

In this chapter, we covered many of the steps needed to mix and master your recording and prepare it for final delivery. We covered what you need to do and get to achieve a good mix, such as quality monitors, headphones, treating your space correctly, and balancing your tracks. We looked at how to adjust faders, panning, automation, and much more. We discussed the basics of using effects and how to apply them. We also covered common equalization techniques to enhance the tonal balance of instruments in your mix. We also covered how you optimize your DAW power by utilizing Auxiliary Input tracks and routing your signal so that you can reduce RTAS overhead.

We covered the process of doing a final mixdown to other components, such as a mastering deck, or other outboard gear such as an analog microphone preamplifier. We covered what the mastering process is and the steps and equipment needed to apply final effects and processing to your work, and prepare a Red Book CD. You are now ready to prepare for final delivery of the work and, to do so, you will need to know how to configure Pro Tools for final exporting of your project or how to finalize your work in a secondary deck such as the Alesis MasterLink or in mastering software.

In the next and final chapter, we will cover delivery. This is the final process of the workflow. In this chapter, we will cover how to import and export data from Pro Tools in all available formats, cover common problems associated when saving and storing sessions, session backup, as well as finalizing your work for distribution and release.

In this chapter

7 Importing and delivery

In this chapter, we cover the final delivery of your session, including the Session Bounce and how to choose from various options. This chapter also covers the steps that you take after final mixdown to create a high-quality CD and also discusses how to properly use iTunes and other importing/exporting tools. In this chapter, we also cover how to prepare for any loss of data. As you will discover, any production facility today that works with digital audio or video files finds hard disk space to be a challenge. Finally, this chapter covers how to store your sessions and software, prepare for the worst, and get back online if any issues do occur via a backup of your system.

7.1 Introduction

Delivery is the final process of the workflow. In this chapter, we cover how to import and export data from Pro Tools in all available formats, cover common problems associated when saving and storing sessions, discuss session backup, and look at how to finalize your work for distribution and release; we take the final mixdown and process it as a file if bounced or send it to a mastering deck for final processing; we cover how to make a final CD of your work for duplication. The final "master" disk will be used to create the rest of the copies (or duplicates) when you mass-distribute your work. We also needed to prepare your work for online streaming and other formats.

You've spent a lot of time creating a great sounding product ... don't ruin it! You can potentially ruin your mastered work by reimporting it into other programs with lower bit rate and sample rate settings, which can remove a lot of your hard work by stripping your file down (and ultimately your sound) and cheapening it or running it completely. In this chapter, we cover how to create the best final master possible. Finally, we cover how to safeguard your work and how to repair common issues with files and sessions. We end this chapter and book with a section on storing and safeguarding your session data.

Note

Pro Tools allows you to export audio regions as files and save sessions in a format compatible with earlier versions of Pro Tools. Be aware that if in Chapter 1 you took the time to configure your hard disks and spent time in Chapter 2 prepping your session and Workspace correctly, you will be able to move session data from disk to disk without issue. If you start to cross over into new file formats and different standards, you will alter your files and quite possibly render them useless. It is recommended to make a backup of your work in case you make mistakes, which could potentially cost you all your time and work.

7.2 Importing options

Pro Tools provides many ways to import audio files and regions into an open session. Audio files can be imported to either a track or to the Region List. Once imported, audio regions may be dragged to tracks from the Region List. Audio files may also be imported into Pro Tools by dragging the files onto the Pro Tools application icon. Many different types of files can be imported by Pro Tools, including AIFF, WAV, SD II, SD I, MP3, AIFL, WMA, MXF, REX 1 and 2 files, ACID files, QuickTime, and Real Audio files. Table 7.1 shows the typical file formats used with Pro Tools 8 LE.

Note

Some files (such as OMF) need the DigiTranslator2 to work within Pro Tools 8 LE. This can be found in the Complete Music Production Toolkit.

Although there are many other formats, Pro Tools 8 LE will know how to use and work with each of these files by default. While working with a final mixdown, when you bounce, import, and export, you will be given multiple format options in which you save your work. Because of this, it's important to understand where you will be sending your work next. For example, if you are finalizing your work for a master, then once you select the file type, the sample rate, and the bit depth, you can apply your settings and get to work. If you decide you want to save your files and session data on a hard disk, you can decide which file types make the most sense if you want to interchange them between other Digidesign or third-party vendor systems.

Audio CDs are formatted at 16 bit with a sample rate of 44.1 kHz. If you need to prepare your formats for video, you may need to change the formats to 24 bit and 48 kHz. The main difference is that work recorded at 48 k has

Table 7.1 Typical file formats used with Pro Tools 8 LE	
AIFF	Audio Interchange File Format (AIFF), commonly used by Apple.
WAV	WAV (or WAVE): the new AES31/Broadcast Wave format is the current standard for all*.wav files. This is sometimes referred to as BWF or broadcast wave file. This type of file is usually a great representation of your work, but file sizes are quite large to store all the audio data required for the representation.
SD II (SD2)	Sound Designer II (and the older SD I (SD1)) format are used primarily by Apple systems.
MP3	MPEG-1 Audio Layer 3, more commonly referred to as MP3. The de facto data-reduced or data-compressed standard used today because of its smaller file size.
WMA	Windows Media Audio (WMA) is an audio-data compression technology developed by Microsoft. Files sometimes seen as WMA, WMV, or ASF.
REX 1 and 2	REX (ReCycle export) is very helpful for working with tempos. REX (and the newer REX2) is an audio loop file format that was developed by Propellerhead.
ACID	Currently up to Acid Pro 7, this is Sony's file type and is used for working with ACID-based DAWs.

a higher sample rate than 44.1 k, giving the recording a slightly higher frequency response. If you change the sampling rate from 48–44.1 k (without doing a proper conversion or rerecording the work from scratch), your work will sound as if it's slower and lower in pitch. In practice, you should always use 44.1 at 16 bit for CD quality and 48 k and 24 bit for video formats, at least until 44.1 is replaced with 88.2 or 96 k. This is because the sound quality advantage will be, if anything, minimal when switching to 48 k.

It is recommended that you record at the sample rate you wish to finalize your work in and convert as few times as possible. Also, if you change a session from one sample rate to another (44.1–48 k as an example), you may ruin your session by speeding it up and raising the pitch slightly. If this happens, you can repair it in your Workspace by opening your Workspace as seen in Figure 7.1 and selecting the session files you want to work with.

Figure 7.1 Viewing a session in the Workspace.

To change the sample rate of a file, simply use the Workspace to add the new sample rate manually. You can now import the session data into a new session configured at the correct sample rate and you may be able to repair your work. However, this does not always work. An example of how you can change a file's information is seen in Figure 7.2.

virgo live.R.wav	Audio File	39.61 MB	WAV (BWF)	44100	
virgo live.L.wav	Audio File	40.86 MB	WAV (BWF)	44100	24 bits
virgo live x.wav	Audio File	39.62 MB	WAV (BWF)	44100	24 bits
vb1.wav	Audio File	23.89 MB	WAV (BWF)	44100	16 bits
Untitled1.mid	MIDI File	1.26 KB	MIDI (Type 0)		

Figure 7.2 Modifying a file's sample rate in the Workspace.

Caution

When importing, Pro Tools converts the files for playback in the current session. Audio files that are of the same type and sample rate of the session are added directly to the session. Audio files that have a different bit depth or sample rate from the session must be converted for use with the current session.

You can also set the quality settings for importing. The quality of the conversion can be set in the Pro Tools Preferences by clicking on the Processing tab and choosing the desired conversion quality from the drop-down menu. In addition, stereo files are converted by Pro Tools into two mono files for use with the session. Figure 7.3 shows the Processing tab with adjustable sample rate conversion quality.

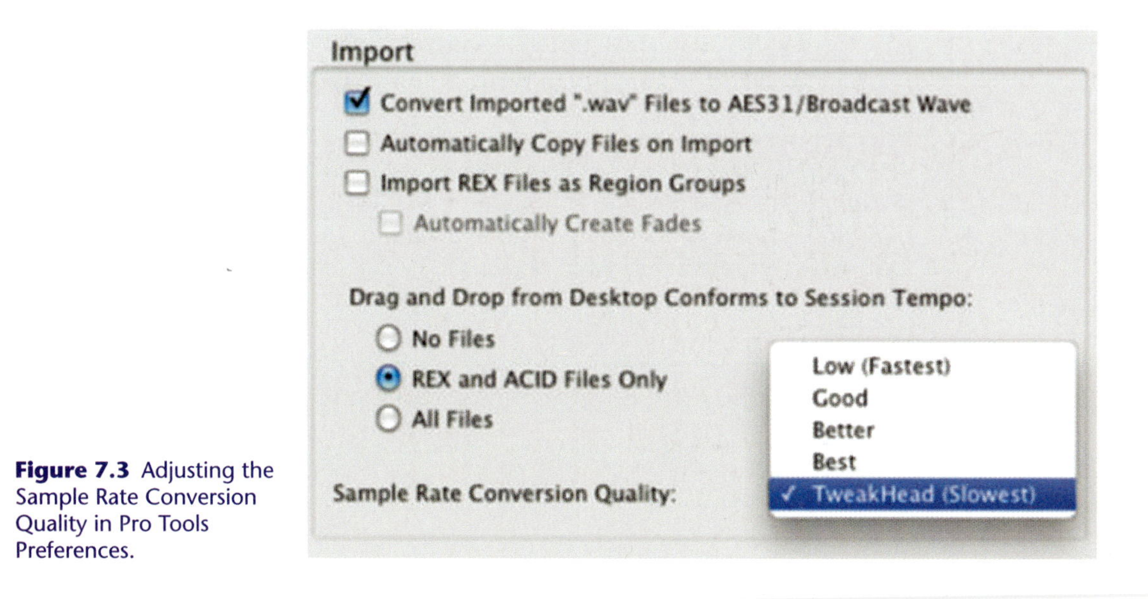

Figure 7.3 Adjusting the Sample Rate Conversion Quality in Pro Tools Preferences.

Note

You can also convert imported WAV files to the new AES31/Broadcast Wave, which is the current standard set for all *.wav files.

Although you may want to select the best quality for your work, be aware that the higher the quality, the longer it takes to import, export, or bounce.

Next, we will look at how to import audio to a track, import audio to a Region List, and import audio using Hardware Inputs (Mbox2 or the 003). To import files to new tracks, click on File, Import, and then Audio to Track. This command will create a new track for each file being imported. Files can also be imported to new tracks by dragging the files from Windows Explorer or Mac Finder to an empty space in the Edit window. Dragging audio files into the Track List will also import the audio into new tracks. The Import Audio window is displayed in Figure 7.4.

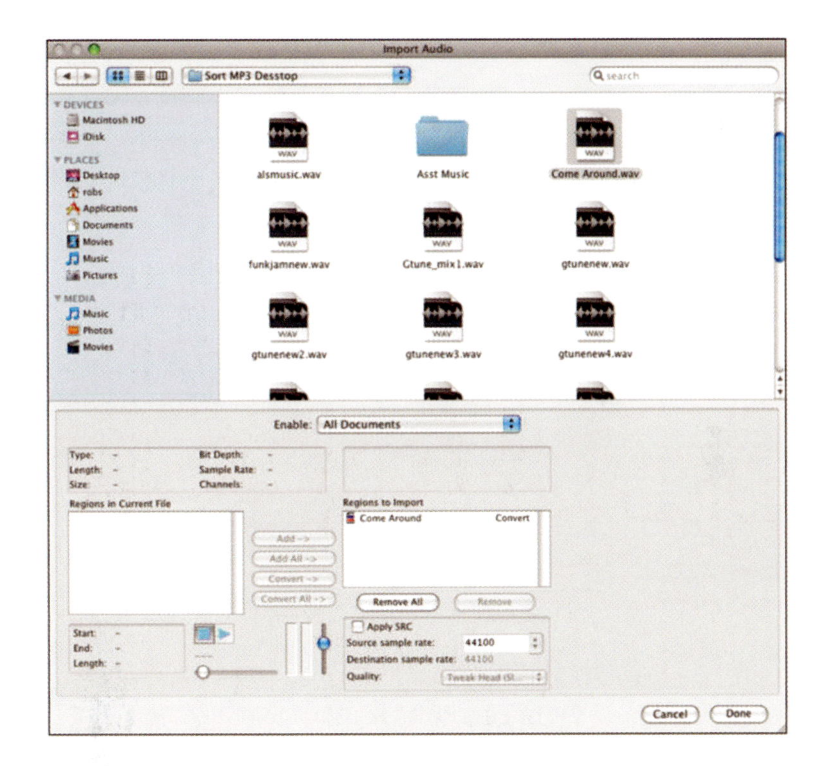

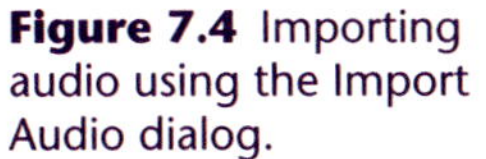

Figure 7.4 Importing audio using the Import Audio dialog.

In this dialog, you can pull any acceptable file format and prepare them to be imported into Pro Tools 8 LE. You can import AIFF, WAV, MP3, SD I, SD II, REX, AAC, and MP4 files, among others.

Notes

Advanced Audio Coding (AAC) is the current contender for replacing MP3 files. AAC is a standard that has many bit-rate configurations and allows for better sound quality than MP3s. Pro Tools cannot import protected AAC or MP4 files with the M4p extension. You should note that these files are protected under the rules of digital rights management or DRM and will not load into your session.

Pro Tools 8 LE supports mixed audio file formats. WAV, SD II, and AIFF files can be used within LE, M-Powered, and HD sessions.

In this example, we have taken a WAV file and selected it, which places it in the Regions to Import section. Here, you can remove the file as well as add more. You can add multiple files at once to be converted in tandem. In Figure 7.5, you can see that the file type is a WAV, its bit depth is 16 bit, and its sample rate is set to 44.1 k.

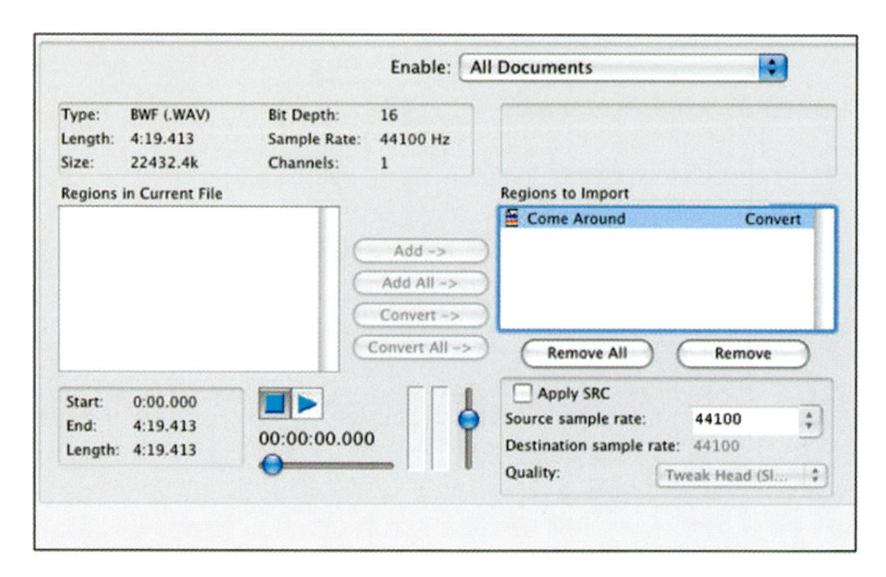

Figure 7.5 Viewing an Imported files configuration in the Import Audio window.

You can review the audio before importing it to check the file and make sure you want to import it for use within Pro Tools 8 LE. In the Import Audio window, audio files may be previewed before importing by clicking on the Play and Stop buttons. The volume may be adjusted by using the vertical slider. The horizontal slider beneath the Play and Stop buttons can be used to fast-forward or rewind through the preview. Figure 7.6 shows the Play and Stop buttons, volume fader, and placement slider.

Once you have checked everything over, you can name and save the file. This will import the file into Pro Tools and create a region that you can add to a track. Alternately, you can create a new track with the saved region already in place. Figure 7.7 shows the two options given on import.

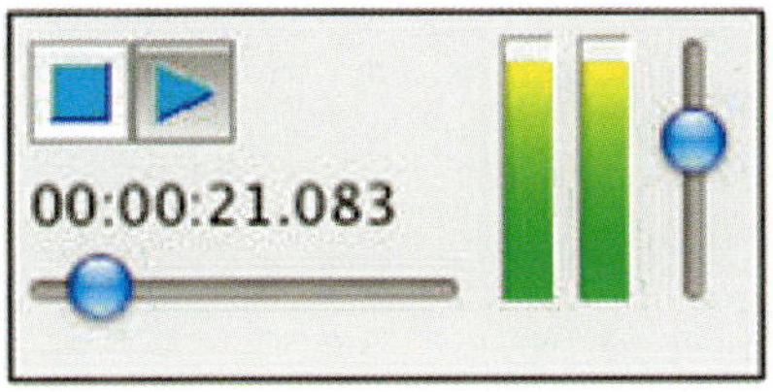

Figure 7.6 Checking a file's output level before import.

Figure 7.7 Creating a New Track or adding data to the Region List.

Note

Tracks from audio CDs can be imported from the Import Audio window or by dragging them into the session. Pro Tools converts the file for use with the current session and saves it to your hard drive.

Files may be imported to the Region List by clicking on File, Import, and then Audio to Region List. Files can also be imported by dragging files from Windows Explorer or Mac Finder to the Region List. Regions can be dragged directly from the Region List into existing tracks.

After files are added, they can be edited like any other region in your Edit window's Region List. Simply import files as needed and you are on your way. Figure 7.8 shows different import options for Pro Tools 8 LE such as Session Data, Audio, MIDI, Video and Region Groups.

Tips

Double-clicking a file in the Import Audio window will automatically add it to the list of Regions to Import.

To quickly import a file from your desktop or Finder, simply drag the file into Pro Tools 8 LE's Edit window. If you drag the file into the Project window, it will either add the track if none are present or add the data to a preexisting track and create a region in the Region List. You can also drag a file directly into the Region List to add it.

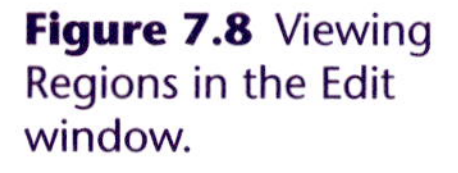

Figure 7.8 Viewing Regions in the Edit window.

A useful import option beyond importing Audio, MIDI, or Video data is the Session Data option in Figure 7.9.

If you choose "Session Data," you will be able to import parts of your session you may need – such as regions and tracks. To import session data, click on Session Data to open the Import Session Data dialog seen in Figure 7.10. Here, you can configure what will be used from the session you are importing, as well as what the destination will be.

Figure 7.9 The Import Audio drop-down menu.

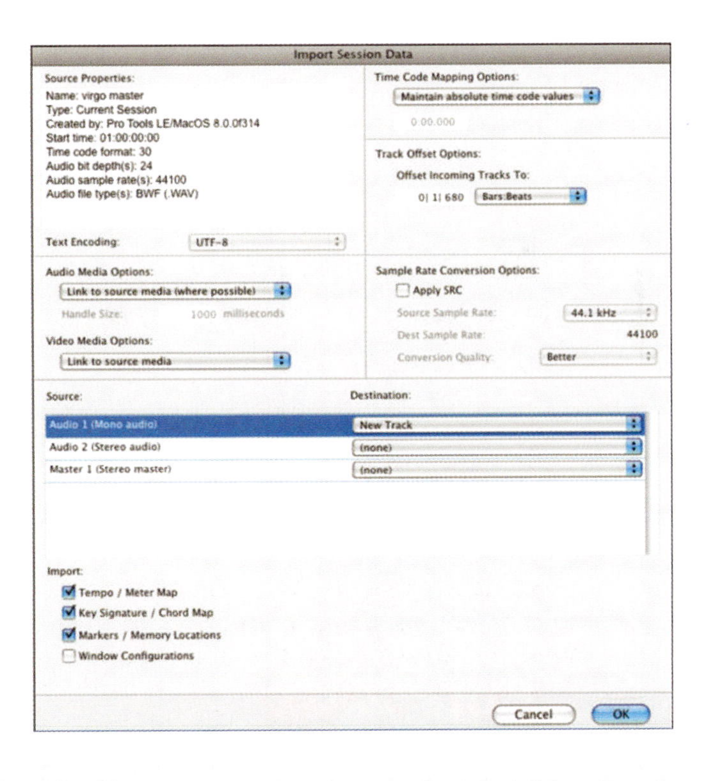

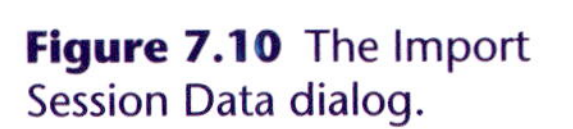
Figure 7.10 The Import Session Data dialog.

The Import Session Data dialog allows you to import tracks from another session into the current session. This dialog shows the properties of the source session, and there are options to copy or link to audio or video tracks.

Note

While importing, you can use Sample Rate Conversion options to apply a new sample rate to your recording. Remember, converting it may change the pitch and speed of the recorded work.

In the Source and Destination sections of the Import Session Data dialog, you can see that Audio 1 (from the session you are importing) is configured to be added as a New Track in the destination session. You can also configure settings for importing a Tempo/Meter Map, Key Signature/Chord Map, Markers/Memory Locations, and Window Configurations, all features covered in Chapter 5, Editing.

Time Code Mapping Options and Track Offset Options are also available in the Import Session Data dialog. Session files created in older versions of Pro Tools can be imported into Pro Tools 8 from older versions of Pro Tools. These older session types are listed in the Open dialog box, seen in Figure 7.11.

To open the session, simply navigate to it and import the tracks needed. Figure 7.12 shows the selection of a new Pro Tools session file that we will import into our current session.

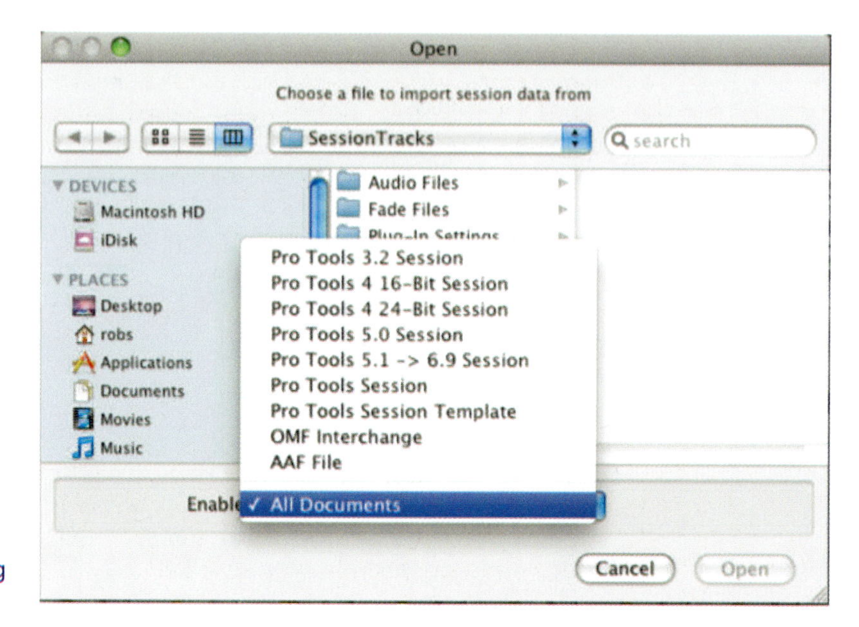

Figure 7.11 Configuring the session type.

Figure 7.12 Importing session data.

Note

In Chapter 5, Editing, we covered the import of audio regions. In this chapter, you will learn how to export audio to a track and export regions as files. You can also send your submix or final mix to a mastering deck, which was covered in Chapter 6, Mixing.

Audio can be imported into a Pro Tools session by recording a source such as a CD, DVD, or DAT player directly into the inputs of your Mbox2 or 003 Rack. The 003 Rack provides the ALT SRC (Alternate Source) input for these types of line level devices. For the Mbox2, connect the audio source to the input channels, create a track, and route the input of the track to the appropriate channel. Record the source material into the Pro Tools session. For the 003 Rack, connect the audio source to the inputs or to the ALT SRC input. Create a track and route the input of the track to the appropriate channel. Record the source material into the Pro Tools session.

Note

The ALT SRC input of the 003 Rack can be used to import audio from another playback device into a Pro Tools session. The ALT SRC To IN 7-8 button on the 003 Rack hardware should be enabled, and the destination track input should be set to Analog 7–8.

As an alternative, you can put a CD or DVD in your computer's CD-ROM or DVD drive, then import the files there into Pro Tools. This is faster than copying the audio in real time and results in no loss in audio quality.

Now that we can import audio into our session without any issues, let's look at final exporting and delivery options.

Video Import, Export, and Bounce

Although Pro Tools needs the production toolkit to give you full access to all tools for working with video, you can import and export video easily. You can import videos by going to the File menu, selecting Import, and then selecting Video. You can also drag and drop from your desktop or Finder if you need to. By selecting Import, you now have the ability to import a video file into Pro Tools. Figure 7.13 shows the Video Import Options dialog. Make sure that you select Import Audio from File if you want to rework the audio in a mono or stereo audio track.

This can be helpful if you want to quickly clean up the audio of a video clip you may want to put online. The difference it makes in the final master is worth the effort, as this gives you the ability to use your entire creative toolset on the audio track. The video track is also imported on its own track as seen in Figure 7.14.

Now you can line up the audio to your video effortlessly without having to export to iMovie, Final Cut, or whatever video-editing suite you work with. You can create a great looking and amazing sounding clip quickly and easily with Pro Tools 8 LE. To further the process for the Web, you can reload the file into iMovie, Final Cut, QuickTime, Windows Media Player, or any other editing tool and change to other formats with streaming features enabled.

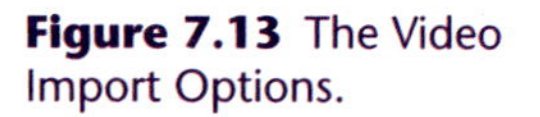

Figure 7.13 The Video Import Options.

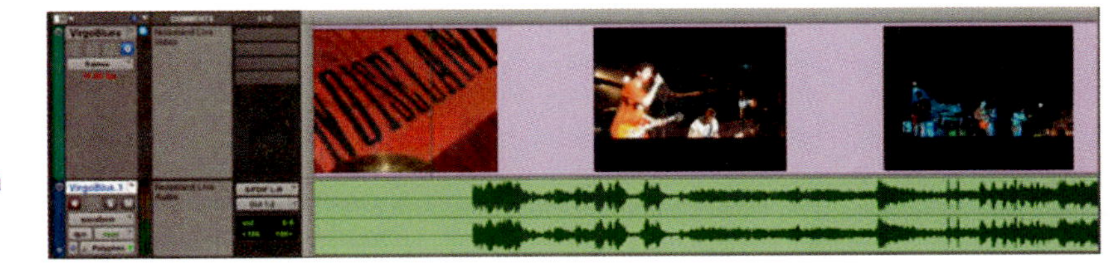

Figure 7.14 Viewing a Video Track in the Edit window.

7.3 Exporting options

If your final mixdown is mastered and ready to export as a *.wav file, do so. There are also a few other export options available with Pro Tools 8 LE that allow you to take session data and save it as a MIDI file or export it to Sibelius. You can also export to Selected Tracks as New AAF/OMF by selecting the File menu and choosing Export. There you can choose to have selected tracks if you have DigiTranslator 2 configured. Audio regions may be exported by selecting regions in the Region List and choosing the Exporting Regions as Files command found in the Region List drop-down menu seen in Figure 7.15.

Find... ⇧⌘F
Clear Find ⇧⌘D
✓ Select Parent in Project Browser
Show ▸
Select ▸
Sort by ▸
Clear... ⇧⌘B
Rename... ⇧⌘R
Auto Rename...
Time Stamp... ⇧⌘M
Compact...
Export Region Definitions... ⇧⌘Y
Export Regions as Files... ⇧⌘K
Export Region Groups...
Recalculate Waveform Overviews
Timeline Drop Order ▸

Figure 7.15 Export Options in the Region List menu.

Region definitions are files that store all the region information for your session. Everything about the session is exported except for the actual audio waveform. This saves a lot of space by saving only the framework of your session and region information. Exporting regions as definitions will export your regions as files but may alter the location in which Pro Tools stores session file data. For example, Figure 7.16 will pop up before you export to alert you that to share region definitions between other sessions, you need to export them as well.

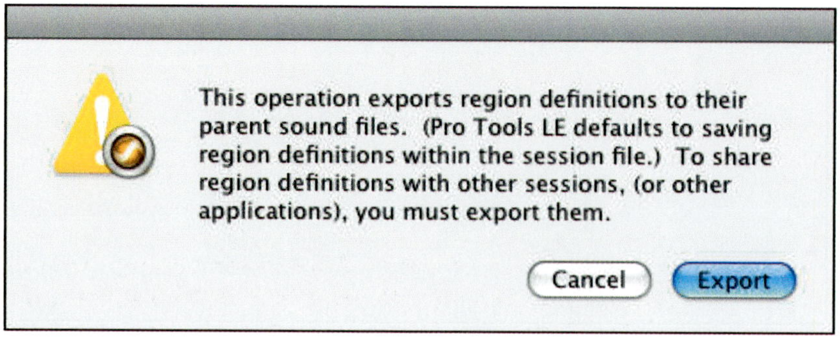

This operation exports region definitions to their parent sound files. (Pro Tools LE defaults to saving region definitions within the session file.) To share region definitions with other sessions, (or other applications), you must export them.

Cancel Export

Figure 7.16 Exporting Regions as definitions.

Caution

Many times, if regions go missing, you can open a session and find that you need to relink the file path in the session to the location of the file that has now been moved. Once found, it will reimport itself and function normally. If it doesn't, you will not be able to use the region and may have to rerecord it.

The Export Selected window that appears after selecting the Export Regions as Files option provides export options for File Type, Mono or Stereo Channel Format, Bit Depth, Sample Rate, and Conversion Quality. The Destination Directory can be selected by clicking on the Choose button. This window also contains options for resolving duplicate file names. To complete the export process, select the desired export options from the Export Selected dialog, select a destination for the files, and click on Export.

Note

Audio files can be exported using Finder or Windows Explorer by moving or copying the files, or by burning them to CD-ROM or DVD-ROM media.

You can export your MIDI data as a file. You can export by going to the File menu and choosing Export. Then select MIDI. You will see the Export MIDI Settings dialog (Fig. 7.17).

Figure 7.17 The Export MIDI Setting dialog.

Once you choose the MIDI File Format and Location Reference, you can click on OK to save the file as a *.mid file. Make sure you remember to select Apply Real-Time Properties if you want to apply any edits to your work in real time, also covered in Chapter 3.

If you want to export your MIDI data as a file in Sibelius, you can do so by going to the File menu, selecting Export, and then choosing Sibelius as seen in Figure 7.18.

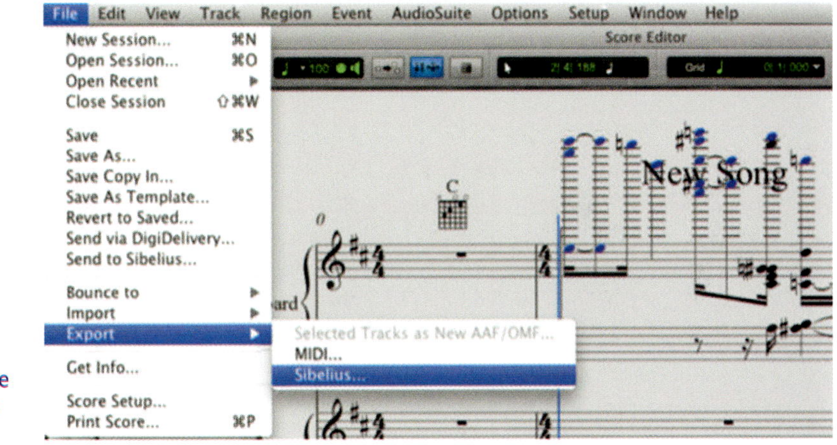

Figure 7.18 Viewing the Export to Sibelius option from the Score Editor.

Tip

The Score Editor must be active to be able to send files to Sibelius, which must be installed on your system. Sibelius 5 file extensions are *.sib and are also exportable back into Pro Tools 8 LE.

7.4 Bouncing options

Before you bounce a final mix to disk, adjust your output levels, check that all your tracks are balanced, make sure a final mixdown is ready to process, and ensure that all tracks are not armed and prepped to go. Then simply go to the File menu and choose Bounce to Disk or Bounce to QuickTime Movie. First, let's cover the audio bounce. The Bounce dialog will open as seen in Figure 7.19.

Once opened, you can make many different settings change that will dramatically affect the sound of your work. These settings can also affect the quality of the file you are trying to create. As we move through this, in the next section, we cover each one and give you pointers as to what to select. First, you should choose the destination path you want to bounce to. By default,

Figure 7.19 The Bounce dialog.

Tip

To quickly pull up the Bounce dialog, you can press the option key, the command key, and then the B key.

Analogs 1 and 2 are selected as a stereo pair. If you choose to send out of your Mbox 2 or 003, refer to our earlier pointers about how to import, export, and bounce from your hardware.

Next, you can configure the File Type. In Figure 7.19, the MP3 option is selected. If you do not have the music production toolkit installed, you will not be able to bounce down in MP3 format. Figure 7.20 shows the error you will get if you do not have the MP3 bundle enabled.

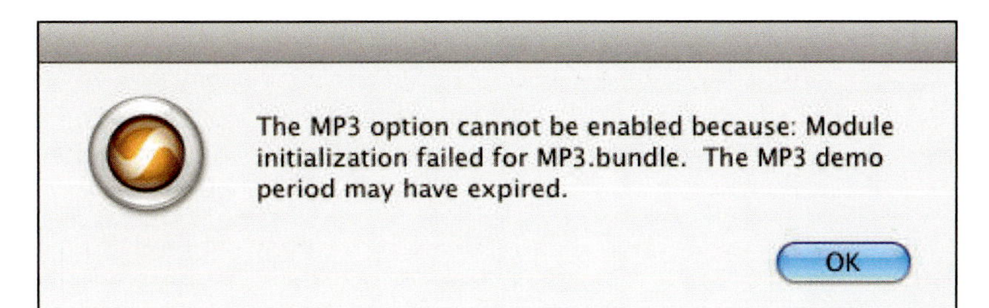

Figure 7.20 Error shown due to MP3 option being disabled.

If you do not choose MP3, you can also select between SD II, WAV (default), AIFF, QuickTime, or MXF. Figure 7.21 shows the different File Types you can select.

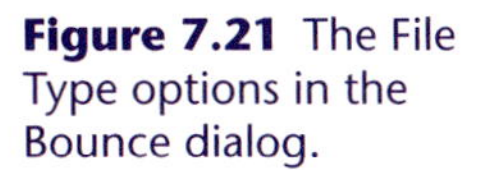

Figure 7.21 The File Type options in the Bounce dialog.

Next, you can adjust the Format, making it Mono (summed), Multiple mono, or Stereo Interleaved (Fig. 7.22).

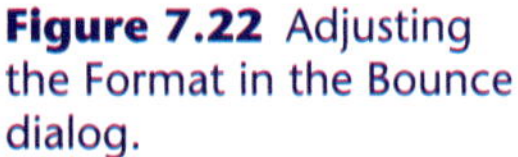

Figure 7.22 Adjusting the Format in the Bounce dialog.

The format you choose will dictate how your file is produced on the hard disk. For example, if you select Mono (summed), you can bounce your final mix-down as a mono file. All tracks will be "summed" to one track (mono). If you choose Multiple mono, it will create a track for each track coming from the source. Since the source is two channels in stereo, your files will bounce as two files in stereo – and be labeled as such. You will have a Left (*.L) file for your left stereo channel and a Right (*.R) file for the right side of the stereo pair. If you select Stereo Interleaved, all bounced tracks will be bounced as a signal stereo "interleave" file.

Bit Depth, which is adjusted directly below Format, is also critical to your bounce. For example, if you select 8 bit, you will be removing 50% of the quality our 16-bit master would have had. That's essentially cutting away 50% of the information needed to represent the file, making it sound very distorted. You want to set a bit depth of 16 bit for CD quality files. If you want to get better quality recordings, then you can select the 24-bit option although it will require you to use (and perhaps purchase) more disk space on your DAW and the 24-bit file will not burn to an audio CD, only to a data CD.

Caution

Setting higher bit depths or sample rates will equate to more disk space being required to store the work – and the demands grow as you record more and more, so make sure that you are aware of your disk space needs and take it into consideration when planning your sessions, sizes, files, and so on.

Sample Rate is configured next. This is the number of times an audio signal is measured every second, measured in kHz. Today's DAWs commonly use 44.1 k for CD-quality work and a 16-bit bit depth. This meets the Red Book standard. 96 k can be used for higher-than-CD quality work. Figure 7.23 shows the adjustment of the Sample Rate where you can select from a predefined set of options or customize your own Sample Rate.

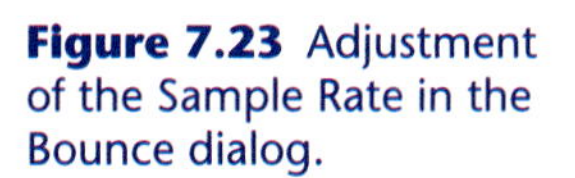

Figure 7.23 Adjustment of the Sample Rate in the Bounce dialog.

Once you have selected the proper Sample Rate, check over your Conversion Quality settings. If you want to get the best conversion quality, use the best option available which is Tweak Head. Low is still decent quality but definitely not the best. When using higher Conversion Quality settings such as Better and Best, you also have to wait longer for the bounce to finish and convert. The Conversion Quality settings can be adjusted in the Conversion Quality portion of the Bounce dialog as seen in Figure 7.24.

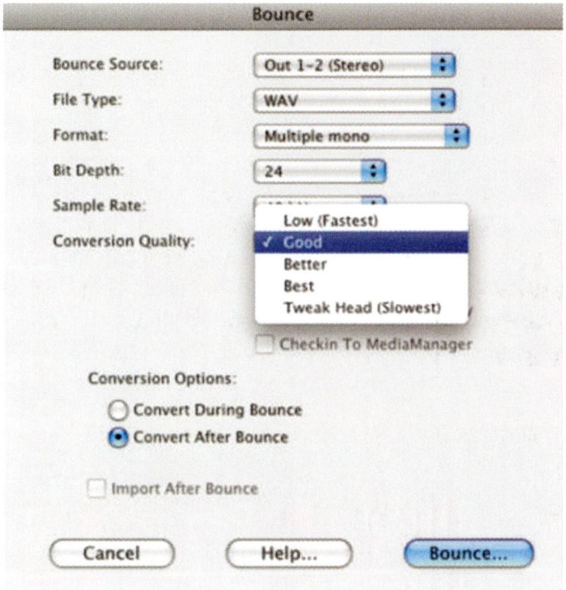

Figure 7.24 Adjusting the Conversion Quality settings in the Bounce dialog.

The Conversion Quality options will also allow you to use the Squeezer, Enforce Avid Compatibility (if enabled), and Checkin to MediaManager. Both of the latter functions will come in to play if you use the Avid Unity client with your setup. The Squeezer is used for high-end 8-bit conversions from a 16-bit final mix.

If you are using lower bit depths, such as 8 bit, you can apply a squeezer tool to apply compression and limiting features before converting your work down to 8 bit. This gives the file a better chance at sounding better when bounced.

Tip

Conversion Options include Convert During Bounce and Convert After Bounce. You can choose to convert during the bounce for a faster bounce, but less quality in the processed file, or you can select to convert it after the bounce, which will give it the highest quality. You can also choose to Import After Bounce if you have configured and are using a submix or a stem.

Once you have selected your options, click on the Bounce button. You can save the bounced file using the Save dialog box in Figure 7.25. You can now select a destination folder for the new audio file, then name and save it.

Figure 7.25 The Save Bounce As dialog.

Tip

Click on the New Folder button on the bottom left hand side of the Save dialog. This way, you can organize your folders in a hierarchy and keep them neat, organized, and easy to find in case of problems.

Once you click Save, you must wait for the bounce to complete and then you can load your file onto a CD, into iTunes, or wherever else you want it. You have to wait for the bounce process to finish – there are only limited things you can do when the bounce process kicks off. For one, you may want to limit the amount of processes on your system as you bounce because the process takes up a lot of CPU and memory. If you find that your bouncing procedure takes too long or times out before completing, you may not have enough resources to complete the work. While bouncing, it's recommended you wait and use the time to perform one last listen of the work to check it over for last-minute mistakes or changes. Figure 7.26 shows the bouncing to disk process, which can take as long as it takes to convert and send your work to the hard disk as a file.

While the bounce process takes place, prepare a CD-R (recordable) to copy your work on. Our next section covers final processing and saving your work

Bouncing to Disk...

Time Remaining: 5:01

(type Command-period or Escape to cancel)

Figure 7.26 Bouncing your work to Disk.

so that you do not lose it in case you have a hard-drive failure or need to reclaim lost data.

Exporting video to QuickTime

As we covered earlier in this chapter, you can import and export video to and from Pro Tools 8 LE. As mentioned earlier, Pro Tools 8 LE needs the production toolkit to give you full access to all tools available to working with video. By going to the File menu, selecting Bounce to and then selecting Video you can create a new QuickTime Movie as seen in Figure 7.27.

To bounce, simply change the settings such as Format, Bit Depth, Sample Rate, and Conversion Quality as necessary (Fig. 7.28).

Once you have completed the bounce, you can check your movie for visual quality. If you are not happy with the quality, you can simply rebounce it again with higher settings, this time hopefully improving your video and/or audio tracks.

Figure 7.27 Bouncing a QuickTime Movie from Pro Tools.

Your final step to closing out the project is either to bounce the file down to your hard disk for further processing or to send the signal to a mastering deck for further processing.

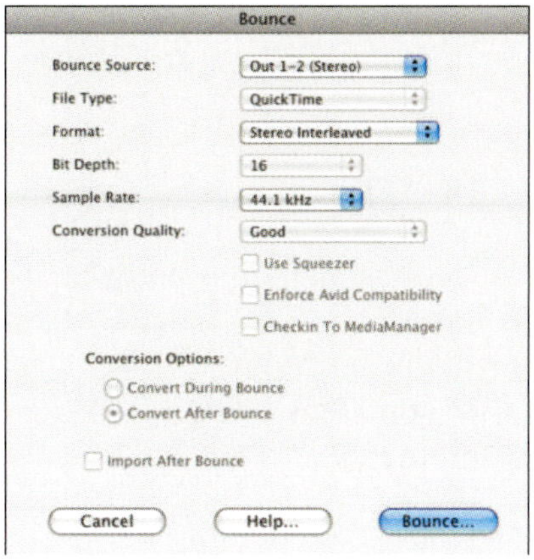

Figure 7.28 The Bounce dialog when bouncing video.

Note

In Chapter 6, Mixing, we covered the use of a mastering deck such as the Alesis MasterLink.

After you have burned your master CD, the next step is to use a duplication system that burns CDs at high speed.

If you bounce a file to your hard disk, your next step would be to create a CD, so you can listen to it.

Tips

Do not import your work to iTunes without configuring it properly. For example, if you want to import your WAV files into iTunes, make sure to configure your iTunes preferences correctly or you may not get the best quality out of your final product. To correct any potential issues in iTunes, simply open the preferences, and then choose Import Settings. Here, you can adjust the file used when importing into iTunes, as well as the quality. With MP3s (for example), you can adjust the quality from 16 to 320 kbps. Use the higher bit rate for better quality imports.

If you are going to upload your files to the Internet, make sure that you review each Web site's policy on uploads. Each will tell you what formats and sizes work best and are allowed. Some sites have restrictions that will impact the quality of your work. Upload the highest quality that you can and make sure you do test mixes in many sources (as covered in Chapter 6, Mixing) so that you get the best possible quality. This means, listen back to your work in the same mediums in which you plan to use them in and test thoroughly.

After you have finalized your work, have your master in hand, and are ready to duplicate, you have one step left – clean up!

7.5 Backup and recovery

As an audio engineer or producer, you will come to the close of every project with not only a CD in hand but also a lot of final cleanup, backup, and shutdown steps to complete. As we covered at the end of the recording phase (in Chapter 4), when you finish your work, you need to power down your DAW correctly. For example, lowering your levels, powering down your monitors, and saving your work are some of the first steps to take once you have completed your session. The next step is to make sure that you leave your DAW ready for the next session if you are working in a studio, where you may have new clients coming in. If no other clients are coming in, then you can backup your sessions and clean up the studio.

It's always recommended that when you work with Pro Tools (or any software for that matter), you get a clean backup of your work. This means that you know that the backup taken has been completed properly so that in the case of data loss, you can quickly get back to work without too many problems or headaches.

Obviously, you want to keep working copies of all your software. Your DAW components, your computer system, Pro Tools itself, plug-ins, and other third-party tools – everything you need to get back and running should be duplicated and kept safe in the unlikely event your computer crashes or Pro Tools itself becomes problematic.

Tip

Software when used over time can (quite frequently) become corrupted. A virus, system crash, or any number of glitches can ruin your day and disable your ability to record. Also, since you are moving large amounts of data across drives in your system, you will likely need to run disk cleanup tools on your hard disks to keep them running smoothly. As we mentioned in Chapter 1, it is important to prepare your hard drives correctly before installing Pro Tools. You can also backup your session data (or have it on an external drive) and reinstall Pro Tools again if you run out of options while trying to repair a Pro Tools-related problem caused by software issues.

Backup of Session Data can be done in many ways. You can have a backup on an external drive. You can mirror a drive to another drive so that a copy is kept on both mirrors. If a drive fails, you have a backup solution ready to go. This is easier than using the backup tools that come with your computer system – most external drive vendors sell their drives with software that can enable backup tools and features if needed.

You can also save your data on a centralized network server or FTP server. This is recommended if you are doing local and centralized backups of your systems to a tape device. You would need to take the tapes to another location or keep them in a fireproof safe to ensure that your protected data remains available in case of disaster.

Finally, it's important to consider off-site backups, especially if you are working with other clients. Many times, if you are working with other bands, they sometimes bring in their own disks and make their own copies and backup. If not, it's recommended that you do just in case. Nothing is worse than spending four weeks on a recording only to lose it forever because of a software problem.

Tip

Off-site backup solutions are not cheap, so it might be worth trying one online. Some online backup solutions offer late-night backups over the network where your session data can be stored and reclaimed easily. You will need to consider how this impacts your Internet line, and if privacy is a concern, you may not feel comfortable with a solution of this kind.

Note

Climate control is extremely important. Hard disks commonly fail due to extreme heat. If all your important data is on the drive, you may lose it. Data recovery options on failed hardware can be expensive. Online backup of sessions might help solve any climate-related dilemmas.

7.6 Summary

In this chapter, we wrapped up the production workflow by exporting your mix and preparing it for duplication by making a CD master or converting it to MP3 to put on the Web, or other formats. Delivery is the final stage in the workflow. From start to finish, you have configured a DAW, upgraded to Pro Tools 8 LE and walked through the recording process for audio and MIDI recording, edited, mixed, mastered, and delivered your work. Whether you wrote a song, are working alone, are producing a band, or are recording a choir, the workflow usually does not change drastically. If anything, you will use a workflow of this sort 90% of the time you do produce or engineer a project. Editing usually does not come before recording and mixing definitely does not come before DAW setup.

This being said, try to remember this workflow process and use it every step of the way in your next recording project. Get your DAW setup, configure a session and prepare it for use, create your tracks, write your song, record your MIDI and audio work, and then prepare it for final processing. Chapter 5 covered the editing process where you were able to further edit your work and enhance it. While learning how to edit, we also looked at the many tools that come with Pro Tools 8 LE such as Beat Detective. In Chapter 6, we learned how to then take our final product and mix it correctly, master it, and further enhance it with effects. We learned the finer details of preparing your work for final delivery and covered many formats and configurable options available to you.

In this chapter, we covered how to import and export data from Pro Tools in all available formats, covered common problems associated when saving and storing sessions, discussed session backup, and looked at how to finalize your work for distribution and release. We took the final mixdown and bounced it as a file or sent it to a mastering deck for final editing and compression. Next are keyboard shortcuts and tips to help increase your productivity at the console. This handy Appendix will give you some insight as to navigating a session quickly while using not only a control surface and a mouse but also the keys of the keyboard.

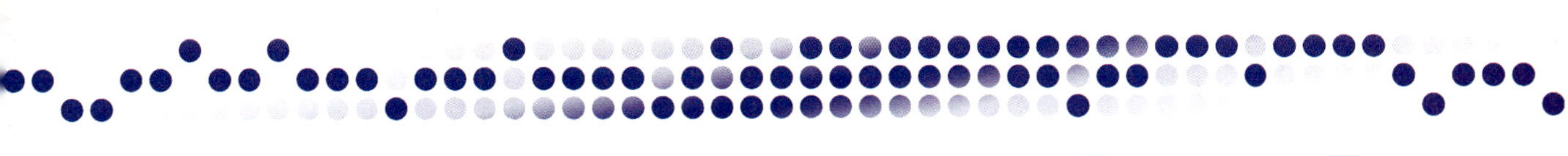

Appendix

The ultimate printout and quick reference guide to mastering the keyboard with Pro Tools 8.

A.1 Using keyboard shortcuts

Whether composing, recording, configuring your session, editing, mixing, or finalizing your work, keyboard shortcuts are available at every phase of the production workflow to help make your time spent at the console easier. Not only do they keep you from editing aerobics, which is the constant back and forth motion you may encounter while at the console, but they also keep you moving through the process by allowing you a way to work with two hands while engineering. In this Appendix, we go over how to speed up your workflow within Pro Tools not only by using shortcuts but also by using multiple methods and tools to help you become even more productive in your recording sessions. You will find many ways to increase productivity by using both your mouse and the keyboard simultaneously. By working both hands, you will not only save time by moving quicker by removing steps but also work twice as fast because you are using two input devices to get things done.

Pro Tools shortcuts make your life easier while working – there is no doubt about it. Imagine having to move the mouse to do everything it is you do within Pro Tools – it's just not realistic when working in a professional environment. To invoke the Automation window for a quick mixing task, you could quickly hit the Command key and then number 4 on the keypad on the right side of your keyboard. To hide it again, simply do the command sequence again. This is definitely very handy, but one would ask how all these commands and shortcuts are memorized. Well, the answer is simple. Anyone benefiting from the key commands is likely using Pro Tools often enough to spend the time memorizing them. Also, the shortcuts and commands are laid out so that you can remember them by associating them in groups. For example, by holding down the Command key again and then hitting the 1, 2, and 3 keys on the keypad open the Transport window, the Session Setup dialog, and the Big Counter window, respectively. Now, as a test, take your finger off the Command key and depress the Option key. Now hit 1, 2, and 3 on your keypad again, and you opened the Time Operations window, the Tempo Operations window, and the Event Operations window, respectively. To have opened all those tools with a mouse, you would have had to make half a dozen menu trips up and down the screen. Now that you see how easy they can be to learn and how quick they make your production move, let's start to look at how you can learn and then work with them into a session.

If you are using a control surface, you can bypass most of the keyboard and mouse usage completely, but it's still required to do many tasks such as typing in names. It's helpful to keep all three nearby. By using all of them simultaneously, you will become highly effective and efficient in just about any task you take on within the workflow.

Note

A control surface can be used to help you handle most tasks easily and quickly without having to use the keyboard. Chapter 1, Introduction, covers the use of control surfaces with Pro Tools systems.

This Appendix covers some of the most common things you will encounter while working within Pro Tools, both on a Leopard or on a Vista system. This Appendix cannot cover everything you can do within Pro Tools on the keyboard, mouse, and/or control surface; a volume of its own would need to be drawn up to encompass all you can do. It does, however, give you a look at the most commonly used shortcuts for making any step of your production workflow quicker, easier, and more comfortable.

Caution

Sessions can be long, and you can spend a lot of time at the console or keyboard as mentioned in Chapter 1, Introduction. Ergonomics are crucial to keeping comfortable and avoiding injury. As silly as it may sound, you can in fact wear yourself out working as an audio engineer through button pushing, fader sliding, and keyboard tapping your way to a finished work. You may find that after 10 years of working this way five nights a week, your back is a little stiff and your neck is sore. Make sure that you sit at your DAW comfortably while working.

Note

In this Appendix, we will discuss and go over many commonly used Mac OS X shortcuts. The end of this Appendix also covers commonly used shortcuts for Windows Vista as well.

A.2 Leopard operating system shortcuts

To learn to use Pro Tools efficiently and help you quickly navigate some of the most common things you will do while working with Pro Tools, let's consider your operating system first. There are a few things that you can do within the Leopard environment that will save you time, help you work within Pro Tools, and ultimately be more productive.

Inevitably something may go wrong with your system during its lifetime, so instead of being stranded in a session, you can quickly reboot your system from a lockup with a single keystroke. You shouldn't be at the mercy of your system, so learning a few of the system's keyboard shortcuts may prove helpful – especially when having a problem. For example, you may have a malfunctioning peripheral. A mouse may be unresponsive, or a keyboard may not be functional. Whatever your dismay, you're only a key command away from getting right back into control. Leopard has many keyboard shortcuts that can help get you out of jams. To gain control of an unresponsive system and shut it down completely, do the following key command:

Control + Option + Command + Eject

This will shut down your system completely and will require you to manually restart it. If you want to regain control of your Mac but only restart it, use a similar key command:

Control + Command + Eject

Performing this step will gracefully shutdown or restart your system – you will not lose any information during this process. Performing both of these steps will quit all your open applications correctly by asking you to save any changes, allowing you to save work that you may have not had a chance to save yet. Once you save your work, the system is either shutdown or restarted.

There may be times where you may need to put your Mac to sleep. The sleep function will save system resources – such as laptop battery life. You can do this by doing the following key command:

Command + Option + Eject

If Pro Tools is the suspected culprit, you can force quit the program. Understand that doing so will result in loss of work. Also know that powering off the unit without closing Pro Tools correctly or saving your work will also result in loss of data or data corruption not only to your session files but also to the operating system itself. To force quit, enter the following key command:

Option + Command + esc

Note

If the computer is not responding and you know that everything you have tried has failed, you can attempt to power the system down by pressing the reset button and then the power button. If all else fails, pulling the plug from the power socket will absolutely do the trick. If this happens (or it ultimately comes down to severing the power from your DAW), you probably had something serious happen to your system, which most likely needs some attention. It is recommended that you look at your system closely and perform system maintenance to check for any problems, as well as take the steps to repair them. If you do not know what you are looking for, have a professional who knows how to check and repair systems to take a look and help you out.

Your system should boot up and be error-free before opening Pro Tools. It helps you to rule out what may be going on while Pro Tools is running, such as errors that may occur from not having enough system resources needed to support Pro Tools 8 LE. Once up at the desktop, you can open Pro Tools and begin your session.

Tip

If you want a change of perspective to take the strain off your eyes during long sessions, you can toggle the inverse of your current view by doing the following key command:

Control + Option + Command + *

(* is the asterisk above the 8 keyboard key)

This will switch the view of your system in a way that may be more helpful to you, as it may help take the strain off your eyes. You can perform the key command again to flip it back to its original settings.

A.3 Pro Tools 8 LE shortcuts (Leopard)

Once you open Pro Tools 8 LE, what task you perform will dictate what keyboard shortcuts are available to you at that time. For example, you may be working within the Edit window and need to start a new session. What would you do? Without shortcuts, you would go to the File menu and then select New Session from the drop-down menu. You will be making a great many sessions, so it will be helpful to know some shortcuts to make your life easier. Table A.1 lists some of the most common shortcuts that you can use to get started quickly.

Table A.1 Keyboard shortcuts	
Create a new session	Command + N
To open a preexisting session	Command + O
To save your work	Command + S
To close your session	Command + Shift + W

Once you have mastered these few key commands, move on to the tools that you find yourself using most of the time while working with Pro Tools. Your toolbox provides you with what you need to navigate the editing process. To navigate through the different editing modes, you can use first four function keys on your keyboard. You can also access your editing tools with these keys as well. Table A.2 lists these shortcuts.

Table A.2 Keyboard shortcuts	
To access Shuffle mode	F1 key
To access Slip mode	F2 key
To access Spot mode	F3 key
To access Grid mode	F4 key
To access the Zoomer tool	F5 key
To access the Trimmer tool	F6 key
To access the Selector tool	F7 key
To access the Grabber tool	F8 key
To access the Scrubber tool	F9 key
To access the Pencil tool	F10 key
Clear the screen of open Pro Tools Windows	F11 key
Record (if a track is record enabled)	F12 key

You can also save time by learning keyboard shortcuts when working with Memory Locations. Memory Locations allow you to mark points on your time-line while recording your session. As an engineer, you will find yourself either

Tip

You can also "right-click" your mouse in the project section of the Edit window to invoke a menu that also helps you access tools and other services. To access the Smart tool, press the F6 + F7 or F7 + F8 keys on your keyboard.

working with Memory Locations often or not at all – it's all a matter of preference. Since many do choose to use it, we will cover it here.

Memory Locations come in extremely handy when you want to make a reference point. You may need to mark an area, so you can quickly find it when needed. As an example, you can make a marker if wanted to name each song. If you are recording multiple songs, you could name each one with a reference point and then access them quickly. You can access stored Memory Locations in the Window menu. Selecting Memory Locations from the drop-down menu produces a dialog where you can find your stored markers. You can see an example of the Memory Locations dialog in Figure A.1.

Figure A.1 Memory Location dialog.

You can store your edit selections as Memory Locations. Memory Locations are also a great way to customize different track views so that you can access them with quick key commands.

To create a Memory Location, press the Enter key. To reset a Memory Location, press Control + click on a Memory Location button. To delete a Memory Location, press the Option + click on a Memory Location button.

Other common edit options available to you while working with the keyboard and the Edit window are listed in Table A.3.

Table A.3 Keyboard shortcuts

To save your current settings	Command + Shift + S
To copy current settings	Command + Shift + C
To paste current settings	Command + Shift + V
To undo current settings	Command + Shift + Z
To quit Pro Tools, save your work	Command + Q

Note

If you did not save your work, you will be asked to, then the session will close and Pro Tools will quit.

Tip

Reference the documentation that came with your Pro Tools system for a full set of keyboard shortcuts for every function within your Pro Tools system.

A.4 Using Vista

When using your Windows Vista system (Business or Ultimate Editions) to run Pro Tools 8 LE, you will want to work quickly and efficiently – not only within Pro Tools 8 LE but also with the base operating system. As you get to know your desktop (and Pro Tools), you will find that learning how to use both hands to navigate Pro Tools while working through your session will help you be more productive.

In this Appendix, we go over not only how to move quickly while working within Pro Tools but also how to get the most out of the operating system that you are working with.

This Appendix cannot cover everything you can do within Pro Tools on the keyboard, but a volume is in the documentation included with your Pro Tools system in both Windows and Apple versions. It's recommended that you print whatever copy you need and keep it on your desk for easy reference. Once you do, you will be glad that you did.

To learn to use Pro Tools efficiently and help you quickly navigate some of the most common things you will do while working with Pro Tools, let's consider the Windows operating system first. Here are some helpful startup and shutdown commands you can use to quickly bring your system back online after a lockup, or some other issue or problem.

Inevitably, something may go wrong with your system during its lifetime, so instead of being stranded in a session, you can quickly reboot your system from a lockup with a single keystroke. You shouldn't be at the mercy of your system, so learning a few of the system's keyboard shortcuts may prove helpful – especially when having a problem. For example, you may have a malfunctioning peripheral. A mouse may be unresponsive or a keyboard may not be functional. Whatever your dismay, you're only a key command away from getting right back into control.

To gain control of an unresponsive system and shut it down completely, do the following key command:

Control + alt + delete

You will be given options as to what you may want to do. You can shut down your system, logout, or pull up the Task Manager by selecting Task List ... which if selected will bring up the Task Manager dialog and open up a plethora of options. For example, you may need to end an offending process or shut down an application completely. Task Manager will help you troubleshoot your system. There are many settings to show you how your system is performing, which is critical to monitor when running Pro Tools 8 LE on Vista. You can see an example of the Task Manager in Figure A.2.

Performing this step will gracefully begin the process of shutting down or restarting your system – you will not lose any information during this process if done correctly. You should save all your work and shut down Pro Tools before opening Task Manager. You will not want to shut down your system before you save your work, you may lose it. Once you save your work, the system can be shut down and the data are safe. Once up at the desktop you can open Pro Tools and begin your session. Get familiar with Task Manager; it's

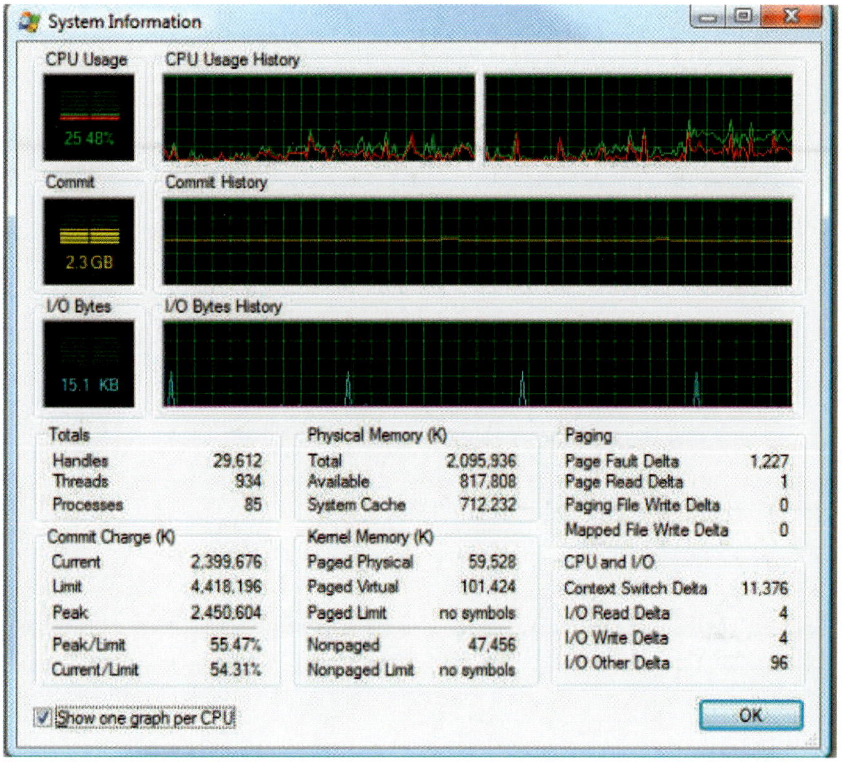

Figure A.2 The Task Manager.

extremely helpful in providing quick solutions and answers to many Windows operating system problems.

A.5 Pro Tools 8 LE shortcuts (Vista)

Once you open Pro Tools 8 LE, what task you perform will dictate what keyboard shortcuts are available to you at that time. For example, you may be working within the Edit window and need to start a new session. What would you do? Without shortcuts, you would go to the File menu and then select New Session from drop-down menu. You will be making a number of sessions, so it's helpful to know some shortcuts to make your life easier. Table A.4 lists helpful ways to invoke a session or close it out.

Once you have mastered these few key commands, move on to the tools you use most of the time with Pro Tools. Your toolbox provides you with what you need to navigate the editing process. To navigate through the different editing modes, you can use first four function keys on your keyboard. Table A.5 shows the keys used to do this.

Table A.4 Keyboard shortcuts	
To create a new session	Control + N
To open a preexisting session	Control + O
To save your work	Control + S
To close your session	Control + Shift + W

Table A.5 Keyboard shortcuts	
To access Shuffle mode	F1 key
To access Slip mode	F2 key
To access Spot mode	F3 key
To access Grid mode	F4 key
To access the Zoomer tool	F5 key
To access the Trimmer tool	F6 key
To access the Selector tool	F7 key
To access the Grabber tool	F8 key
To access the Scrubber tool	F9 key
To access the Pencil tool	F10 key
Clear the screen of open Pro Tools Windows	F11 key
Record (if a track is record enabled)	F12 key

You can also save time by learning keyboard shortcuts when working with Memory Locations. Memory Locations allow you to mark points on your timeline while recording your session. As an engineer, you will find yourself either working with Memory Locations often or not at all – it's all a matter of preference. Since many do choose to use it, we will cover it here.

Memory Locations come in extremely handy when you want to make a reference point. You may need to mark an area, so you can quickly find it when needed. As an example, you can make a marker to name each song or to mark the start of each verse and chorus within a song. If you are recording multiple songs, you could name each one with a reference point and then access them quickly.

You can store your edit selections as Memory Locations. Memory Locations are also a great way to customize different track views that you can configure so that you can access them with quick key commands. To create a Memory Location, press the Enter key on your keyboard. To reset a Memory Location, press Control + click on a Memory Location button. To delete a Memory Location, press Alt + click on a Memory Location button.

Other common edit options available to you while working with the keyboard and the Edit window are in Table A.6.

Table A.6 Keyboard shortcuts

To save your current settings	Control + Shift + S
To copy current settings	Control + Shift + C
To paste current settings	Control + Shift + V
To undo current settings	Control + Shift + Z

Always remember that the more shortcuts you learn, the more productive you will be while working with Pro Tools 8 LE whether on a Vista or Leopard system.

Tip

For a full list of keyboard shortcuts for both platforms, please visit Digidesign's Web site at http://www.digidesign.com or look through the electronic documentation set that comes with your Pro Tools system. A copy of both systems' keyboard shortcuts are neatly organized in an Adobe PDF for your review. This comes with the Pro Tools systems in both Windows and Apple versions. It's recommended that you print out whatever copy you need and keep it on your desk for easy reference. You will be glad that you did.

Index